THE
FORCES OF MATTER

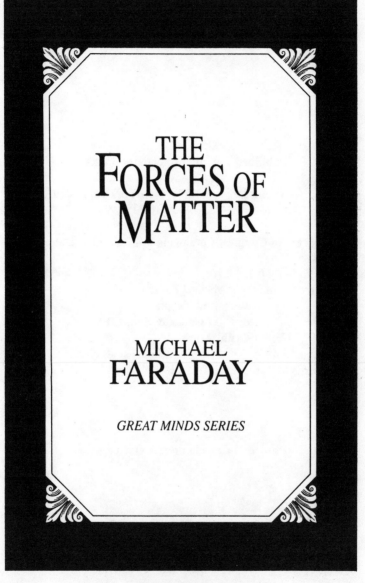

THE
FORCES OF
MATTER

MICHAEL
FARADAY

GREAT MINDS SERIES

PROMETHEUS BOOKS
Buffalo, New York

Published 1993 by Prometheus Books

59 John Glenn Drive, Buffalo, New York 14228-2197,
716-837-2475. FAX: 716-835-6901.

Library of Congress Cataloging-in-Publication Data

Faraday, Michael, 1791–1867.
   The forces of matter / Michael Faraday.
      p.   cm.   (Great minds series)
   Originally published: P. F. Collier & Son, 1910.
   ISBN 0-87975-811-2 (pbk.)
   1. Gravitation. 2. Cohesion. 3. Chemical affinity.
4. Magnetism. 5. Electricity. I. Title. II. Series.
QC178.F37   1993
531.1—dc20   92-41252
                                                    CIP

Printed in Canada on acid-free paper.

Also Available in Prometheus's Great Minds Paperback Series

Charles Darwin
The Origin of Species

Galileo Galilei
*Dialogues Concerning
Two New Sciences*

Edward Gibbon
On Christianity

Ernst Haeckel
The Riddle of the Universe

Herodotus
The History

Julian Huxley
Evolutionary Humanism

Thomas Henry Huxley
*Agnosticism and
Christianity and
Other Essays*

Ernest Renan
The Life of Jesus

Adam Smith
Wealth of Nations

Andrew D. White
*A History of the Warfare
of Science with Theology
in Christendom*

For a complete list of titles in Prometheus's Great Books
in Philosophy and Great Minds Series, see the order form
at the back of this volume.

MICHAEL FARADAY was born at Newington Butts, near London, on September 22, 1791. His parents had migrated from Yorkshire to London, where his father worked as a blacksmith. In his teens, Faraday became an apprentice to a bookbinder, a position he held for eight years. During this time he read extensively on his own, and developed an interest in the experimental study of nature. In 1813, at the age of twenty-one, Faraday was hired by Sir Humphrey Davy as his laboratory assistant at the Royal Institution; from 1813 to 1815, Faraday traveled extensively with Sir Humphrey on the Continent, where he met some of the leading scientists of the day. In 1816, Faraday published his first scientific paper. In 1825, he was appointed director of the laboratory at the Royal Institution, and in 1833 he was made Fullerian Professor for life.

Proclaimed the greatest experimental philosopher of his age, Faraday made numerous contributions to chemistry and physics. Among his most important were the discoveries of magneto-electric induction, electrochemical decomposition, the magnetization of light, and diamegnetism. But the esteemed professor was also a devoted popularizer of science, and delighted in promoting the fascination of the sciences to the young. The following lectures offer to the general reader classic examples of this famed researcher's clarity and forcefulness in explaining various scientific phenomena.

Michael Faraday died at Hampton Court on August 25, 1867.

Faraday's published works include: *Chemical Manipulation* (1827), *Experimental Researches in Electricity* (1839–1855), *Experimental Researches in Chemistry and Physics* (1859), and *Lectures on the Chemical History of a Candle* (1861).

Contents

Lecture I: The Force of Gravitation   5

Lecture II: Gravitation—Cohesion   21

Lecture III: Cohesion—Chemical Affinity   36

Lecture IV: Chemical Affinity—Heat   50

Lecture V: Magnetism—Electricity   62

Lecture VI: The Correlation of the Physical Forces   75

3

THE FORCES OF MATTER

By Michael Faraday

LECTURE I

THE FORCE OF GRAVITATION

IT grieves me much to think that I may have been a cause of disturbance to your Christmas arrangements, for nothing is more satisfactory to my mind than to perform what I undertake; but such things are not always left to our own power, and we must submit to circumstances as they are appointed. I will to-day do my best, and will ask you to bear with me if I am unable to give more than a few words; and, as a substitute, I will endeavor to make the *illustrations* of the sense I try to express as full as possible; and if we find by the end of this lecture that we may be justified in continuing them, thinking that next week our power shall be greater, why, then, with submission to you, we will take such course as you may think fit, either to go on or discontinue them; and although I now feel much weakened by the pressure of the illness (a mere cold) upon me, both in facility of expression and clearness of thought, I shall here claim, as I always have done on these occasions, the right of addressing myself to the younger members of the audience; and for this purpose, therefore, unfitted as it may seem for an elderly, infirm man to do so, I will return to second childhood, and become, as it were, young again among the young.

Let us now consider, for a little while, how wonderfully we stand upon this world. Here it is we are born, bred, and live, and yet we view these things with an almost entire absence of wonder to ourselves respecting the way in which all this happens. So small, indeed, is our wonder, that we are never taken by surprise; and I do think that, to a young

5

person of ten, fifteen, or twenty years of age, perhaps the first sight of a cataract or a mountain would occasion him more surprise than he had ever felt concerning the means of his own existence; how he came here; how he lives; by what means he stands upright; and through what means he moves about from place to place. Hence, we come into this world, we live, and depart from it, without our thoughts being called specifically to consider how all this takes place; and were it not for the exertions of some few inquiring minds, who have looked *into* these things, and ascertained the very beautiful laws and conditions by which we *do* live and stand upon the earth, we should hardly be aware that there was any thing wonderful in it. These inquiries, which have occupied philosophers from the earliest days, when they first began to find out the laws by which we grow, and exist, and enjoy ourselves, up to the present time, have shown us that all this was effected in consequence of the existence of certain *forces,* or *abilities* to do things, or *powers,* that are so common that nothing can be more so; for nothing is commoner than the wonderful powers by which we are enabled to stand upright: they are essential to our existence every moment.

It is my purpose to-day to make you acquainted with some of these powers; not the vital ones, but some of the more elementary, and what we call *physical* powers; and, in the outset, what can I do to bring to your minds a notion of neither more nor less than that which I mean by the word *power* or *force?* Suppose I take this sheet of paper, and place it upright on one edge, resting against a support before me (as the roughest possible illustration of something to be disturbed), and suppose I then pull this piece of string which is attached to it. I pull the paper over. I have therefore brought into use a *power* of doing so—the *power* of my hand carried on through this string in a way which is very remarkable when we come to analyze it; and it is by means of these powers conjointly (for there are several powers here employed) that I pull the paper over. Again, if I give it a push upon the other side, I bring into play a *power,* but a very different exertion of power from the former; or, if I take now this bit of shell-lac [a stick of shell-lac about 12 inches long and 1 1-2 in diameter], and rub it with flannel, and hold it an inch or so in front of the upper part of this upright sheet, the paper is immediately moved towards the shell-lac, and by now drawing the latter away, the paper falls over without having been touched by any thing. You see, in the first illustration I produced an effect than which nothing could be commoner; I pull it over now, not by means of that string or the pull of my hand, but by some action in this shell-lac. The shell-lac, therefore, has a *power* wherewith it acts upon the sheet of paper; and, as an illustration of the exercise of another kind of power, I might use gunpowder with which to throw it over.

Now I want you to endeavor to comprehend that when I am speaking of a *power or force,* I am speaking of that which I used just now to pull over this piece of paper. I will not embarrass you at present with the *name* of that power, but it is clear there was a *something* in the shell-lac which acted by attraction, and pulled the paper over; this, then, is one of those things which we call *power,* or *force;* and you will now be able to recognize it as such in whatever form I show it to you. We are not to suppose that there are so very many different powers; on the contrary, it is wonderful to think how few are the powers by which all the phenomena of nature are governed. There is an illustration of another kind of power in that lamp; *there* is a power of heat—a power of doing something, but not the same power as that which pulled the paper over; and so, by degrees, we find that there are certain other powers (not many) in the various bodies around us; and thus, beginning with the simplest experiments of pushing and pulling, I shall gradually proceed to distinguish these powers one from the other, and compare the way in which they combine together. This world upon which we stand (and we have not much need to travel out of the world for illustrations of our subject; but the mind of man is not confined like the matter of his body, and thus he may and does travel outward, for wherever his sight can pierce, there his observations can penetrate) is pretty nearly a round globe, having its surface disposed in a manner of which this terrestrial globe by my side is a rough model; so much is land and so much is water, and by looking at it here we see in a sort of map or picture how the world is formed upon its surface. Then, when we come to examine farther, I refer you to this sectional diagram of the geological strata of the earth, in which there is a more elaborate view of what is beneath the surface of our globe. And, when we come to dig into or examine it (as man does for his own instruction and advantage, in a variety of ways), we see that it is made up of different kinds of matter, subject to a very few powers; and all disposed in this strange and wonderful way, which gives to man a history—and such a history—as to what there is in those veins, in those rocks, the ores, the water-springs, the atmosphere around, and all varieties of material substances, held together by means of *forces* in one great mass, 8,000 miles in diameter, that the mind is overwhelmed in contemplation of the wonderful history related by these strata (some of which are fine and thin like sheets of paper), all formed in succession by the forces of which I have spoken.

I now shall try to help your attention to what I may pay by directing to-day our thoughts to one kind of power. You see what I mean by the term *matter*—any of these things that I can lay hold of with the hand, or in a bag (for I may take hold of the air by inclosing it in a bag)— they are all portions of matter with which we have to deal at present, generally or particularly, as I may require to illustrate my subject. Here is the sort of matter which we call *water*—it is there ice [pointing to a block of ice upon the table], *there* water—[pointing to the water boiling in a flask]—*here* vapor—you see it issuing out from the top [of the flask]. Do not suppose that that ice and that water are two entirely different things, or that the steam rising in bubbles and ascending in vapor *there* is absolutely different from the fluid water: it may be different in some particulars, having reference to the *amounts* of power which it contains; but it is the same, nevertheless, as the great ocean of water around our globe, and I employ it here for the sake of illustration, because if we look into it we shall find that it supplies us with examples of all the powers to

which I shall have to refer. For instance, here is water—
it is heavy; but let us examine it with regard to the *amount*
of its heaviness or its gravity. I have before me a little
glass vessel and scales [nearly equipoised scales, one of
which contained a half-pint glass vessel], and the glass
vessel is at present the lighter of the two; but if I now take
some water and pour it in, you see that that side of the
scales immediately goes down; that shows you (using common language, which I will not suppose for the present
you have hitherto applied very strictly) that it is *heavy,*
and if I put this additional weight into the opposite scale,
I should not wonder if this vessel would hold water enough
to weigh *it* down. [The lecturer poured more water into
the jar, which again went down.] Why do I hold the
bottle *above* the vessel to pour the water into it? You will
say, because experience has taught me that it is necessary.
I do it for a better reason—because it is a law of nature
that the water should fall toward the earth, and therefore
the very means which I use to cause the water to enter
the vessel are those which will carry the whole body of
water down. That power is what we call *gravity,* and you
see *there* [pointing to the scales] a good deal of water
gravitating toward the earth. Now *here* [exhibiting a small
piece of platinum([1])] is another thing which gravitates
toward the earth as much as the whole of that water. See
what a little there is of it; that little thing is heavier than
so much water [placing the metal in opposite scales to the
water]. What a wonderful thing it is to see that it re-
quires so much water as *that* [a half-pint vessel full] to
fall toward the earth, compared with the little mass of
substance I have *here!* And again, if I take this metal
[a bar of aluminium([2]) about eight times the bulk of the
platinum], we find the water will balance that as well as it
did the platinum; so that we get, even in the very outset,
an example of what we want to understand by the words
forces or *powers.*

I have spoken of water, and first of all of its property of
falling downward: you know very well how the oceans sur-

[1] *Platinum,* with one exception the heaviest body known, is 21½ times
heavier than water.
[2] *Aluminium* is 2½ times heavier than water.

round the globe—how they fall round the surface, giving roundness to it, clothing it like a garment; but, besides that, there are other properties of water. *Here,* for instance, is some quicklime, and if I add some water to it, you will find another power and property in the water.([3])   It is now very hot; it is steaming up; and I could perhaps light phosphorus or a lucifer-match with it. Now that could not happen without a *force* in the water to produce the result; but that force is entirely distinct from its power of falling to the earth. Again, here is another substance [some anhydrous sulphate of copper([4])] which will illustrate another kind of power. [The lecturer here poured some water over the white sulphate of copper, which immediately became blue, evolving considerable heat at the same time.] Here is the same water with a substance which heats nearly as much as the lime does, but see how differently. So great indeed is this heat in the case of lime, that it is sufficient sometimes (as you see here) to set wood on fire; and this explains what we have sometimes heard, of barges laden with quicklime taking fire in the middle of the river, in consequence of this power of heat brought into play by a leakage of the water into the barge. You see how strangely different subjects for our consideration arise when we come to think over these various matters—the power of heat evolved by acting upon lime with water, and the power which water has of turning this salt of copper from white to blue.

I want you now to understand the nature of the most simple exertion of this power of matter called *weight* or *gravity.* Bodies are heavy; you saw that in the case of water when I placed it in the balance. Here I have what we call a *weight* [an iron half cwt.]—a thing called a weight because in it the exercise of that power of pressing downward is especially used for the purposes of weighing; and I have also one of these little inflated India-rubber bladders, which are

---

[3] *Power or property in water.* This power—the heat by which the water is kept in a *fluid* state—is said, under ordinary circumstances, to be *latent* or *insensible.* When, however, the water changes its form, and, by uniting with the lime or sulphate of copper, becomes *solid,* the heat which retained it in a liquid state is evolved.

[4] *Anhydrous sulphate of copper:* sulphate of copper deprived of its water of crystallization. To obtain it the blue sulphate is calcined in an earthen crucible.

very beautiful although very common (most beautiful things are common), and I am going to put the weight upon it, to give you a sort of illustration of the downward pressure of the iron, and of the power which the air possesses of resisting that pressure; it may burst, but we must try to avoid that. [During the last few observations the lecturer had succeeded in placing the half cwt. in a state of quiescence upon the inflated India-rubber ball, which consequently assumed a shape very much resembling a flat cheese with round edges.] There you see a bubble of air bearing half a hundred weight, and you must conceive for yourselves what a wonderful *power* there must be to pull this weight downward, to sink it thus in the ball of air.

Let me now give you another illustration of this power. You know what a pendulum is. I have one here (FIG. 1), and if I set it swinging, it will continue to swing to and fro.

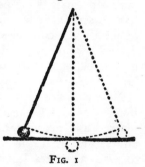

FIG. 1

Now I wonder whether you can tell me why that body oscillates to and fro—that pendulum bob, as it is sometimes called. Observe, if I hold the straight stick horizontally, as high as the position of the ball at the two ends of its journey, you see that the ball is in a higher position at the two extremities than it is when in the middle. Starting from one end of the stick, the ball falls toward the centre, and then rising again to the opposite end, it constantly tries to fall to the lowest point, swinging and vibrating most beautifully, and with wonderful properties in other respects—the time of its vibration, and so on—but concerning which we will not now trouble ourselves.

If a gold leaf, or piece of thread, or any other substance were hung where this ball is, it would swing to and fro in the same manner, and in the same time too. Do not be startled at this statement; I repeat, in the same manner and in the same time, and you will see by-and-by how this is. Now that power which caused the water to descend in the balance—which made the iron weight press upon and flatten the bubble of air—which caused the swinging to and fro of the pendulum, that power is entirely due to the attraction which there is between the falling body and the earth. Let us be slow and careful to comprehend this. It is not that the earth has any *particular* attraction toward bodies which fall to it, but, that *all* these bodies possess an attraction every one toward the other. It is not that the earth has any special power which these balls themselves have not; for just as much power as the earth has to attract these two balls [dropping two ivory balls], just so much power have they in proportion to their bulks to draw themselves one to the other; and the only reason why they fall so quickly to the earth is owing to its greater size. Now if I were to place these two balls near together, I should not be able, by the most delicate arrangement of apparatus, to make you, or myself, sensible that these balls did attract one another; and yet we know that such is the case, because if, instead of taking a small ivory ball, we take a mountain, and put a ball like this near it, we find that, owing to the vast size of the mountain as compared with the billiard ball, the latter is drawn slightly toward it, showing clearly that an attraction *does* exist, just as it did between the shell-lac which I rubbed and the piece of paper which was overturned by it.

Now it is not very easy to make these things quite clear at the outset and I must take care not to leave anything unexplained as I proceed, and, therefore, I must make you clearly understand that all bodies are attracted to the earth, or, to use a more learned term, *gravitate*. You will not mind my using this word, for when I say that this penny-piece *gravitates,* I mean nothing more nor less than that it falls toward the earth, and, if not intercepted, it would go on falling, falling, until it arrived at what we call the *centre of gravity* of the earth, which I will explain to you by-and-by.

I want you to understand that this property of gravitation is never lost; that every substance possesses it; that there is never any change in the quantity of it; and, first of all, I will take as illustration a piece of marble. Now this marble has weight, as you will see if I put it in these scales; it weighs the balance down, and if I take it off, the balance goes back again and resumes its equilibrium. I can decompose this marble and change it in the same manner as I can change ice into water and water into steam. I can convert a part of it into *its own* steam easily, and show you that this steam from the marble has the property of remaining in the same place at common temperatures, which *water*-steam has not. If I add a little liquid to the marble and decompose it (⁵), I get that which you see—[the lecturer here put several lumps of marble into a glass jar, and poured water and then acid over them; the carbonic acid immediately commenced to escape with considerable effervescence] —the appearance of boiling, which is only the separation of one part of the marble from another. Now this [marble] steam, and that [water] steam, and all other steams, *gravitate* just like any other substance does; they all are attracted the one toward the other, and all fall toward the earth, and what I want you to see is that *this* steam gravitates. I have here (FIG. 2) a large vessel placed upon a balance, and the

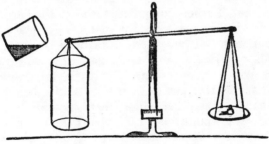

FIG. 2

moment I pour this steam into it you see that the steam gravitates. Just watch the index, and see whether it tilts over or not. [The lecturer here poured the carbonic acid out of the glass in which it was being generated into the vessel suspended on the balance, when the gravitation of the carbonic acid was at once apparent.] Look how it is going

⁵ *Add a little liquid to the marble and decompose it.* Marble is composed of *carbonic acid* and *lime,* and, in chemical language, is called *carbonate of lime.* When sulphuric acid is added to it, the carbonic acid is set free, and the sulphuric acid unites with the lime to form sulphate of lime.

Carbonic acid, under ordinary circumstances, is a colorless invisible gas, about half as heavy again as air. Dr. Faraday first showed that under great pressure it could be obtained in a liquid state. Thilorier, a French chemist, afterward found that it could be solidified.

down. How pretty that is! I poured nothing in but the
invisible steam, or vapor, or gas which came from the marble, but you see that part of the marble, although it has
taken the shape of air, still gravitates as it did before. Now
will it weigh down that bit of paper? [placing a piece of
paper in the opposite scale]. Yes, more than that; it nearly
weighs down this bit of paper [placing another piece of
paper in]. And thus you see that *other* forms of matter
besides solids and liquids tend to fall to the earth; and, therefore, you will accept from me the fact that *all* things gravitate, whatever may be their form or condition. Now *here* is
another chemical test which is very readily applied. [Some
of the carbonic acid was poured from one vessel into another,
and its presence in the latter shown by introducing into it a
lighted taper, which was immediately extinguished.] You
see from this result also that it gravitates. All these experiments show you that, tried by the balance, tried by pouring
like water from one vessel to another, this steam, or vapor,
or gas is, like all other things, attracted to the earth.

There is another point I want in the next place to draw
your attention to. I have here a quantity of shot; each of
these falls separately, and each has its own gravitating
power, as you perceive when I let them fall loosely on a
sheet of paper. If I put them into a bottle, I collect them
together as one mass, and philosophers have discovered that
there is a certain point in the middle of the whole collection
of shots that may be considered as the *one point* in which
all their *gravitating power* is centred, and that point they
call the *centre of gravity;* it is not at all a bad name, and
rather a short one—the centre of gravity. Now suppose I
take a sheet of pasteboard, or any other thing easily dealt
with, and run a bradawl through it at one corner, A
(FIG. 3), and Mr. Anderson holds that up in his hand before
us, and I then take a piece of thread and an ivory ball,
and hang that upon the awl, then the centre of gravity of both
the pasteboard and the ball and string are as near as they
can get to the centre of the earth; that is to say, the
whole of the attracting power of the earth is, as it were,
centred in a single point of the cardboard, and this point
is exactly below the point of suspension. All I have to do,

therefore, is to draw a line, A B, corresponding with the string, and we shall find that the centre of gravity is somewhere in that line. But where? To find that out, all we

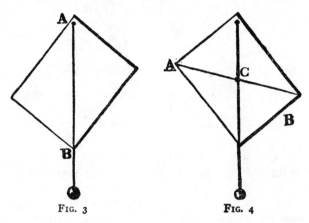

Fig. 3   Fig. 4

have to do is to take another place for the awl (FIG. 4), hang the plumb-line, and make the same experiment, and there [at the point C] is the centre of gravity,—there where the two lines which I have traced cross each other; and if I take that pasteboard and make a hole with the bradawl through it at that point, you will see that it will be supported in any position in which it may be placed. Now, knowing that, what do I do when I try to stand upon one leg? Do you not see that I push myself over to the left side, and quietly take up the right leg, and thus bring some central point in my body over this left leg? What is that point which I throw over? You will know at once that it is the *centre of gravity*—that point in me where the whole gravitating force of my body is centred, and which I thus bring in a line over my foot.

Here is a toy I happened to see the other day, which will, I think, serve to illustrate our subject very well. That toy *ought* to lie something in this manner (FIG. 5), and would do so if it were uniform in substance; but you see it does not; it will get up again. And now philosophy comes to our aid, and I am perfectly sure, without looking inside the figure, that there is some arrangement by which the centre of gravity is at the lowest point when the image is standing upright; and we may be certain, when I am tilting it over

FIG. 5   FIG. 6

(see FIG. 6), that I am lifting up the centre of gravity (*a*), and raising it from the earth. All this is effected by putting a piece of lead inside the lower part of the image, and making the base of large curvature, and there you have the whole secret. But what will happen if I try to make the

figure stand upon a sharp point? You observe I must get that point exactly under the centre of gravity, or it will fall over thus [endeavoring unsuccessfully to balance it]; and this, you see, is a difficult matter; I can not make it stand steadily; but if I embarrass this poor old lady with a world of trouble, and hang this wire with bullets at each end about her neck, it is very evident that, owing to there being those balls of lead hanging down on either side, in addition to the lead inside, I have lowered the centre of

FIG. 7

gravity, and now she will stand upon this point (FIG. 7), and, what is more, she proves the truth of our philosophy by standing sideways.

I remember an experiment which puzzled me very much

when a boy. I read it in a conjuring book, and this was how the problem was put to us: "How," as the book said, "how to hang a pail of water, by means of a stick, upon the side of a table" (Fig. 8). Now I have here a table, a piece of stick, and a pail, and the proposition is, how can that pail be hung to the edge of this table? It is to be done, and can you at all anticipate what arrangement I shall make to enable me to succeed? Why this. I take a stick, and put it in the pail between the bottom and the horizontal piece

FIG. 8   FIG. 9

of wood, and thus give it a stiff handle, and there it is; and, what is more, the more water I put into the pail, the better it will hang. It is very true that before I quite succeeded I had the misfortune to push the bottoms of several pails out; but here it is hanging firmly (Fig. 9), and you now see how you can hang up the pail in the way which the conjuring books require.

Again, if you are really so inclined (and I do hope all of you are), you will find a great deal of philosophy in this [holding up a cork and a pointed thin stick about a foot long]. Do not refer to your toy-books, and say you have seen that before. Answer me rather, if I ask you, have you *understood* it before? It is an experiment which appeared very wonderful to me when I was a boy. I used to take a piece of cork (and I remember I thought at first that it was very important that it should be cut out in the shape of a man, but by degrees I got rid of that idea), and the problem was to balance it on the point of a stick. Now you will see I have only to place two sharp-pointed sticks one on each side, and give it wings, thus, and you will find this beautiful condition fulfilled.

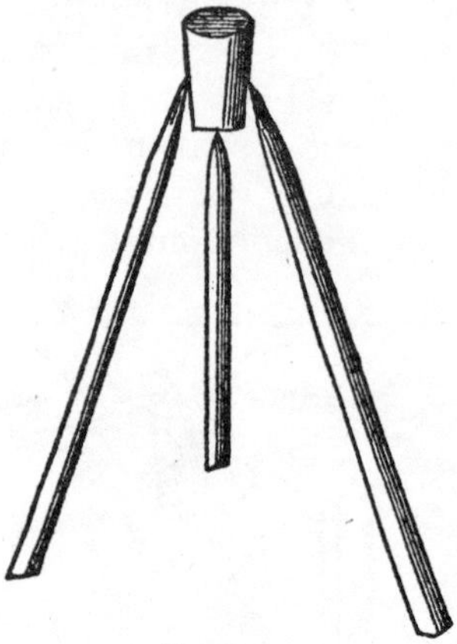

We come now to another point. All bodies, whether heavy or light, fall to the earth by this force which we call gravity. By observation, moreover, we see that bodies do not occupy the same time in falling; I think you will be able to see that this piece of paper and that ivory ball fall with different velocities to the table [dropping them]; and if, again, I take a feather and an ivory ball, and let them fall, you see they reach the table or earth at different times; that is to say, the ball faster than the feather. Now that should not be so, for all bodies do fall equally fast to the earth. There are one or two beautiful points included in that statement. First of all, it is manifest that an ounce, or a pound, or a ton, or a thousand tons, all fall equally fast, no one faster than another: here are two balls of lead, a very light one and a very heavy one, and you perceive they both fall to the earth in the same time. Now if I were to put into a little bag a number of these balls sufficient to make up a bulk equal to the large one, they would also fall in the same time; for if an avalanche fall from the mountains, the rocks, snow, and ice, together falling toward the earth, fall with the same velocity, whatever be their size.

FIG. 10

I can not take a better illustration of this than that of gold leaf, because it brings before us the reason of this apparent difference in the time of the fall. Here is a piece of gold leaf. Now if I take a lump of gold and this gold leaf, and let them fall through the air together, you see that the lump of gold—the sovereign or coin—will fall much faster than the gold leaf. But why? They are both gold, whether sovereign or gold leaf. Why should they not fall to the

earth with the same quickness? *They would do so,* but
that the air around our globe interferes very much where
we have the piece of gold so extended and enlarged as to
offer much obstruction on falling through it. I will, however, show you that gold leaf *does* fall as fast when the
resistance of the air is excluded; for if I take a piece of
gold leaf and hang it in the centre of a bottle so that the
gold, and the bottle, and the air within shall all have an
equal chance of falling, then the gold leaf will fall as
fast as anything else. And if I suspend the bottle containing the gold leaf to a string, and set it oscillating like a
pendulum, I may make it vibrate as hard as I please and
the gold leaf will not be disturbed, but will swing as steadily
as a piece of iron would do; and I might even swing it
round my head with any degree of force, and it would remain undisturbed. Or I can try another kind of experiment:
if I raise the gold leaf in this way [pulling the bottle up to
the ceiling of the theatre by means of a cord and pulley,
and then suddenly letting it fall within a few inches of
the lecture table], and allow it then to fall from the ceiling
downward (I will put something beneath to catch it, supposing I should be *maladroit*), you will perceive that the
gold leaf is not in the least disturbed. The resistance of
the air having been avoided, the glass bottle and gold
leaf all fall exactly in the same time.

Here is another illustration: I have hung a piece of gold
leaf in the upper part of this long glass vessel, and I have
the means by a little arrangement at the top, of letting
the gold leaf loose. Before we let it loose we will remove
the air by means of an air-pump, and, while that is being
done, let me show you another experiment of the same
kind. Take a penny-piece, or a half crown, and a round
piece of paper a trifle smaller in diameter than the coin,
and try them side by side to see whether they fall at the
same time [dropping them]. You see they do not—the
penny-piece goes down first. But, now place this paper
flat on the top of the coin, so that it shall not meet with
any resistance from the air, and upon *then* dropping them
you see they *do* both fall in the same time [exhibiting the
effect]. I dare say, if I were to put this piece of gold

leaf, instead of the paper, on the coin, it would do as well. It is very difficult to lay the gold leaf so flat that the air shall not get under it and lift it up in falling, and I am rather doubtful as to the success of this, because the gold leaf is puckery, but will risk the experiment. There they go together! [letting them fall] and you see at once that they both reach the table at the same moment.

We have now pumped the air out of the vessel, and you will perceive that the gold leaf will fall as quickly in this vacuum as the coin does in the air. I am now going to let it loose, and you must watch to see how rapidly it falls. There! [letting the gold loose] there it is, falling as gold should fall.

I am sorry to see our time for parting is drawing so near. As we proceed, I intend to write upon the board behind me certain words, so as to recall to your minds what we have already examined; and I put the word FORCES as a heading, and I will then add beneath the names of the special forces according to the order in which we consider them; and although I fear that I have not sufficiently pointed out to you the more important circumstances connected with the force of GRAVITATION, especially the law which governs its attraction (for which, I think, I must take up a little time at our next meeting), still I will put that word on the board, and hope you will now remember that we have in some degree considered the *force of gravitation*—that force which causes all bodies to attract each other when they are at sensible distances apart, and tends to draw them together.

LECTURE II

GRAVITATION—COHESION

DO me the favor to pay me as much attention as you did at our last meeting, and I shall not repent of that which I have proposed to undertake. It will be impossible for us to consider the Laws of Nature, and what they effect, unless we now and then give our sole attention, so as to obtain a clear idea upon the subject. Give me now that attention, and then I trust we shall not part without our knowing something about those laws, and the manner in which they act. You recollect, upon the last occasion, I explained that all bodies attracted each other, and that this power we called *gravitation*. I told you that when we brought these two bodies [two equal-sized ivory balls suspended by threads] near together, they attracted each other, and that we might suppose that the whole power of this attraction was exerted between their respective centres of gravity; and, furthermore, you learned from me that if, instead of a small ball I took a larger one, like *that* [changing one of the balls for a much larger one], there was much more of this attraction exerted; or, if I made this ball larger and larger, until, if it were possible, it became as large as the Earth itself—or I might take the Earth itself as the large ball —that *then* the attraction would become so powerful as to cause them to rush together in this manner [dropping the ivory ball]. You sit *there* upright, and I stand upright *here,* because we keep our centres of gravity properly balanced with respect to the earth; and I need not tell you that on the other side of this world the people are standing and moving about with their feet toward our feet, in a reversed position as compared with us, and all by means of this power of gravitation to the centre of the earth.

I must not, however, leave the subject of gravitation without telling you something about its laws and regularity; and, first as regards its power with respect to the distance that bodies are apart. If I take one of these balls and place it within an inch of the other, they attract each other with a certain power. If I hold it at a greater distance off, they attract with less power; and if I hold it at a greater distance still, their attraction is still less. Now this fact is of the greatest consequence; for, knowing this law, philosophers have discovered most wonderful things. You know that there is a planet, Uranus, revolving round the sun with us, but eighteen hundred millions of miles off, and because there is another planet as far off as three thousand millions of miles, this law of attraction, or gravitation, still holds good, and philosophers actually discovered this latter planet, Neptune, by reason of the effects of its attraction at this overwhelming distance. Now I want you clearly to understand what this law is. They say (and they are right) that two bodies attract each other *inversely as the square of the distance*—a sad jumble of words until you understand them; but I think we shall soon comprehend what this law is, and what is the meaning of the "inverse square of the distance."

I have here (FIG. 11) a lamp, A, shining most intensely upon this disc, B, C, D, and this light acts as a sun by which

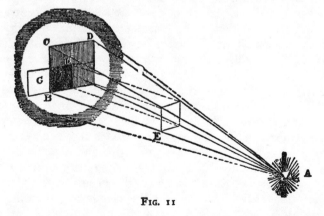

FIG. 11

I can get a shadow from this little screen B F (merely a square piece of card), which, as you know, when I place it close to the large screen, just shadows as much of it as is exactly equal to its own size; but now let me take this card, E, which is equal to the other one in size, and place it midway between the lamp and the screen; now look at the size of the shadow B D—it is four times the original size. Here, then, comes the "inverse square of the distance." This distance, A E, is *one*, and that distance, A B, is *two*, but that size E being *one*, this size B D of shadow is *four* instead of *two*, which is the *square* of the distance, and, if I put the screen at one-third of the distance from the lamp, the shadow on the large screen would be *nine* times the size. Again, if I hold this screen *here*, at B F, a certain amount of light falls on it; and if I hold it nearer the lamp at E, *more* light shines upon it. And you see at once how much —exactly the quantity which I have shut off from the part of this screen, B D, now in shadow; moreover, you see that if I put a single screen here, at G, by the side of the shadow, it can only receive *one-fourth* of the proportion of light which is obstructed. That, then, is what is meant by the *inverse* of the square of the distance. This screen E is the brightest because it is the nearest, and there is the whole secret of this curious expression, *inversely as the square of the distance*. Now if you can not perfectly recollect this when you go home, get a candle and throw a shadow of something—your profile, if you like—on the wall and then recede or advance, and you will find that your shadow is exactly in proportion to the *square* of the distance you are off the wall; and then, if you consider how much light shines on you at one distance, and how much at another, you get the inverse accordingly. So it is as regards the attraction of these two balls; they attract according to the square of the distance, inversely. I want you to try and remember these words, and then you will be able to go into all the calculations of astronomers as to the planets and other bodies, and tell why they move so fast, and why they go *round* the sun without falling into it and be prepared to enter upon many other interesting inquiries of the like nature.

Let us now leave this subject which I have written upon the board under the word FORCE—GRAVITATION—and go a step farther. All bodies attract each other at sensible distances. I showed you the electric attraction on the last occasion (though I did not call it so); that attracts at a distance; and in order to make our progress a little more gradual, suppose I take a few iron particles [dropping some small fragments of iron on the table]. There! I have already told you that in all cases where bodies fall it is the *particles* that are attracted. You may consider these, then, as separate particles magnified, so as to be evident to your sight; they are loose from each other— they all gravitate—they all fall to the earth—for the force of gravitation *never* fails. Now I have here a centre of power which I will not name at present, and when these particles are placed upon it, see what an attraction they have for each other.

Here I have an arch of iron filings (FIG. 12) regularly built up like an iron bridge, because I have put them within

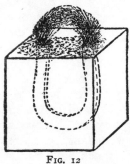

FIG. 12

a sphere of action which will cause them to attract each other. See! I could let a mouse run through it; and yet, if I try to do the same thing with them *here* [on the table], they do not attract each other at all. It is *that* [the magnet] which makes them hold together. Now just as these iron particles hold together in the form of an elliptical bridge, so do the different particles of iron which constitute this nail hold together and make it one. And here is a bar of iron; why, it is only because the different parts of *this* iron are so wrought as to keep close together by the attraction *between* the particles that it is held together in one mass. It is kept together, in fact, merely by the attraction of one particle to another, and that is the point I want now to illustrate. If I take a piece of flint, and strike it with a hammer, and break it thus [breaking off a piece of the flint], I have done nothing more than separate the particles which compose these two pieces so far apart that their attraction is too weak to cause them to hold together, and it is only for that reason that there are now two pieces in the place of one. I will show you an experiment to prove that this attraction does still exist in those particles; for here is a piece of glass (for what was true of the flint and the bar of iron is true of the piece of glass, and is true of every other solid—they are all held together in the lump by the attraction between their parts), and I can show you the attraction between its separate particles; for if I take these portions of glass which I have reduced to very fine powder, you see that I can actually build them up into a solid wall by pressure between two flat surfaces. The power which I thus have of building up this wall is due to the attraction of the particles forming, as it were, the cement which holds them together; and so in this case, where I have taken no very great pains to bring the particles together, you see perhaps a couple of ounces of finely pounded glass standing as an upright wall: is not this attraction most wonderful? *That* bar of iron one inch square has such power of attraction in its particles—giving to it such strength—that it will hold up twenty tons' weight before the little set of particles in the small space equal to one division across which it can be pulled apart will separate. In this manner suspension bridges and chains are held together by the attraction of their particles, and I am going to make an experiment which will show how strong is this attraction of the particles. [The lecturer here placed his foot on a loop of wire fastened to a support above, and swung with his whole weight resting upon it for some moments.] You see, while hanging here, all my weight is supported by these little particles of the wire, just as in pantomimes they sometimes suspend gentlemen and damsels.

How can we make this attraction of the particles a little more simple? There are many things which, if brought together properly, will show this attraction. Here is a boy's experiment (and I like a boy's experiment). Get a tobacco-pipe, fill it with lead, melt it, and then pour it out upon a stone, and thus get a clean piece of lead (this is a better plan than scraping it; scraping alters the condition of the surface of the lead). I have here some pieces of lead which I melted this morning for the sake of making them clean. Now these pieces of lead hang together by the attraction of their particles, and if I press these two separate pieces close together, so as to bring their particles within the sphere of attraction, you will see how soon they become one. I have merely to give them a good squeeze, and draw the upper piece slightly round at the same time, and here they are as one, and all the bending and twisting I can give them will not separate them again; I have joined the lead together, not with solder, but simply by means of the attraction of the particles.

This, however, is not the best way of bringing those particles together; we have many better plans than that; and I will show you 'one that will do very well for juvenile experiments. There is some alum crystallized very beautifully by nature (for all things are far more beautiful in their natural than their artificial form), and here I have some of the same alum broken into fine powder. In it I have destroyed that force of which I have placed the name on this board—COHESION, or the attraction exerted between the particles of bodies to hold them together. Now I am going to show you that if we take this powdered alum and some hot water, and mix them together, I shall dissolve the alum; all the particles will be separated by the water far more completely than they are here in the powder; but then, being in the water, they will have the opportunity as it cools (for that is the condition which favors their coalescence) of uniting together again and forming one mass (^{6}).

Now, having brought the alum into solution, I will pour it into this glass basin, and you will, to-morrow, find that these particles of alum which I have put into the water, and so separated that they are no longer solid, will, as the water cools, come together and cohere, and by to-morrow morning we shall have a great deal of the alum crystallized out—that is to say, come back to the solid form. [The lecturer here poured a little of the hot solution of alum into the glass dish, and when the latter had thus been made warm, the remainder of the solution was added.] I am now doing that

---

^{6} *Crystallization of alum.* The solution must be saturated—that is, it must contain as much alum as can possibly be dissolved. In making the solution, it is best to add powdered alum to hot water as long as it dissolves; and when no more is taken up, allow the solution to stand a few minutes, and then pour it off from the dirt and undissolved alum.

which I advise you to do if you use a glass vessel, namely, warming it slowly and gradually; and in repeating this experiment, do as I do—pour the liquid out gently, leaving all the dirt behind in the basin; and remember that the more carefully and quietly you make this experiment at home, the better the crystals. To-morrow you will see the particles of alum drawn together; and if I put two pieces of coke in some part of the solution (the coke ought first to be washed very clean, and dried), you will find to-morrow that we shall have a beautiful crystallization over the coke, making it exactly resemble a natural mineral.

Now how curiously our ideas expand by watching these conditions of the attraction of cohesion! how many new phenomena it gives us beyond those of the attraction of gravitation! See how it gives us great strength. The things we deal with in building up the structures on the earth are of strength—we use iron, stone, and other things of great strength; and only think that all those structures you have about you—think of the *Great Eastern,* if you please, which is of such size and power as to be almost more than man can manage—are the result of this power of cohesion and attraction.

I have here a body in which I believe you will see a change taking place in its condition of cohesion at the moment it is made. It is at first yellow; it then becomes a fine crimson red. Just watch when I pour these two liquids together—both colorless as water. [The lecturer here mixed together solutions of perchloride of mercury and iodide of potassium, when a yellow precipitate of biniodide of mercury fell down, which almost immediately became crimson red.] Now there is a substance which is very beautiful, but see how it is changing color. It was reddish-yellow at first, but it has now become red ([7]). I have previously prepared a little of this red substance, which you see formed

---

[7] *Red precipitate of biniodide of mercury.* A little care is necessary to obtain this precipitate. The solution of iodide of potassium should be added to the solution of perchloride of mercury (corrosive sublimate) very gradually. The red precipitate which first falls is redissolved when the liquid is stirred: when a little more of the iodide of potassium is added a pale red precipitate is formed, which, on the farther addition of the iodide, changes into the brilliant scarlet biniodide of mercury. If too much iodide of potassium is added, the scarlet precipitate disappears, and a colorless solution is left.

in the liquid, and have put some of it upon paper [exhibiting several sheets of paper coated with scarlet biniodide of mercury] ([8]). There it is—the same substance spread upon paper; and there, too, is the same substance; and here is some more of it [exhibiting a piece of paper as large as the other sheets, but having only very little red color on it, the greater part being yellow]—a *little* more of it, you will say. Do not be mistaken; there is as much upon the surface of one of these pieces of paper as upon the other. What you see yellow is the same thing as the red body, only the attraction of cohesion is in a certain degree changed, for I will take this red body, and apply heat to it (you may perhaps see a little smoke arise, but that is of no consequence); and if you look at it it will first of all darken—but see how it is becoming yellow. I have now made it all yellow, and, what is more, it will remain so; but if I take any hard substance, and rub the yellow part with it, it will immediately go back again to the red condition [exhibiting the experiment]. There it is. You see the red is not *put back,* but *brought back* by the change in the substance. Now [warming it over the spirit lamp] here it is becoming yellow again, and that is all because its attraction of cohesion is changed. And what will you say to me when I tell you that this piece of common charcoal is just the same thing, only differently coalesced, as the diamonds which you wear? (I have put a specimen outside of a piece of straw which was charred in a particular way—it is just like black lead.) Now this charred straw, this charcoal, and these diamonds, are all of them the same substance, changed but in their properties as respects the force of cohesion.

Here is a piece of glass [producing a piece of plate-glass about two inches square]. (I shall want this afterward to look to and examine its internal condition), and here is some of the same sort of glass differing only in its power of cohesion, because while yet melted it had been

---

[8] *Paper coated with scarlet biniodide of mercury.* In order to fix the biniodide on paper, it must be mixed with a little weak gum water, and then spread over the paper, which must be dried without heat.

Biniodide of mercury is said to be *dimorphous;* that is, is able to assume two different forms.

dropped into cold water [exhibiting a "Prince Rupert's drop,"([9]) (Fig. 13)], and if I take one of these little tear-like pieces and break off ever so little from the point, the whole will at once burst and fall to pieces. I will now break off a piece of this. [The lecturer nipped off a small piece from the end of one of Rupert's drops, whereupon the

Fig. 13   Fig. 14

whole immediately fell to pieces.] There! you see the solid glass has suddenly become powder, and more than that, it has knocked a hole in the glass vessel in which it was held. I can show the effect better in this bottle of water, and it is very likely the whole bottle will go. [A 6-oz. vial was filled with water, and a Rupert's drop placed in it with the point of the tail just projecting out; upon breaking the tip off, the drop burst, and the shock, being transmitted through the water to the sides of the bottle, shattered the latter to pieces.]

Here is another form of the same kind of experiment. I have here some more glass which has not been annealed [showing some thick glass vessels([10]) (Fig. 14)], and if I take one of these glass vessels and drop a piece of pounded glass into it (or I will take some of these small pieces of rock crystal; they have the advantage of being harder than glass), and so make the least scratch upon the inside, the whole bottle will break to pieces—it can not hold together. [The lecturer here dropped a small fragment of rock crystal into one of these glass vessels, when the bottom im-

[9] *"Prince Rupert's Drops."* These are made by pouring drops of melted green glass into cold water. They were not, as is commonly supposed, invented by Prince Rupert, but were first brought to England by him in 1660. They excited a great deal of curiosity, and were considered "a kind of miracle in nature."

[10] *Thick glass vessels.* They are called *Proofs* or *Bologna phials.*

mediately came out and fell upon the plate.]   There! it
goes through, just as it would through a sieve.

Now I have shown you these things for the purpose of
bringing your minds to see that bodies are not merely
held together by this power of cohesion, but that they are
held together in very curious ways.   And suppose I take
some things that are held together by this force, and examine them more minutely.   I will first take a bit of glass,
and if I give it a blow with a hammer I shall just break
it to pieces.   You saw how it was in the case of the flint
when I broke the piece off; a piece of a similar kind would
come off, just as you would expect; and if I were to break
it up still more, it would be, as you have seen, simply
a collection of small particles of no definite shape or form.
But supposing I take some other thing—this stone, for
instance (Fig. 15) [taking a piece of mica([11])], and if I

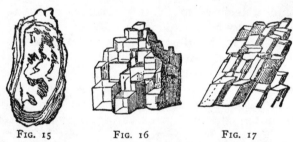

Fig. 15   Fig. 16   Fig. 17

hammer this stone I may batter it a great deal before I can
break it up.   I may even bend it without breaking it—that
is to say, I may bend it in *one particular direction* without
breaking it much, although I feel in my hands that I am
doing it some injury.   But now, if I take it by the edges,
I find that it breaks up into leaf after leaf in a most extraordinary manner.   Why should it break up like that?
Not because all stones do, or all crystals; for there is some
salt (Fig. 16)—you know what common salt is([12]); here
is a piece of this salt, which by natural circumstances has
had its particles so brought together that they have been

[11] *Mica.* A silicate of alumina and magnesia. It has a bright metallic
lustre; hence its name, from *mico,* to shine.
[12] *Common salt* or chloride of sodium crystallizes in the form of solid
cubes, which, aggregated together, form a mass, which may be broken up
into the separate cubes.

allowed free opportunity of combining or coalescing, and you shall see what happens if I take this piece of salt and break it. It does not break as flint did, or as the mica did, but with a clean sharp angle and exact surfaces, beautiful and glittering as diamonds [breaking it by gentle blows with a hammer]; there is a square prism which I may break up into a square cube. You see these fragments are all square; one side may be longer than the other, but they will only split up so as to form square or oblong pieces with cubical sides. Now I go a little farther, and I find another stone (FIG. 17) [Iceland or calc-spar] ([13]) which I may break in a similar way, but *not* with the same result. Here is a piece which I have broken off, and you see there are plain surfaces perfectly regular with respect to each other, but it is not cubical—it is what we call a rhomboid. It still breaks in three directions most beautifully and regularly with polished surfaces, but with *sloping* sides, not like the salt. Why not? It is very manifest that this is owing to the attraction of the particles one for the other being less in the direction in which they give way than in other directions. I have on the table before me a number of little bits of calcareous spar, and I recommend each of you to take a piece home, and then you can take a knife and try to divide it in the direction of any of the surfaces already existing. You will be able to do it at once; but if you try to cut it *across* the crystals, you can not; by hammering you may bruise and break it up, but you can only divide it into these beautiful little rhomboids.

Now I want you to understand a little more how this is, and for this purpose I am going to use the electric light again. You see we can not look into the *middle* of a body like this piece of glass. We perceive the outside form and the inside form, and we look *through* it, but we can not well find out how these forms become so, and I want you, therefore, to take a lesson in the way in which we use a ray of light for the purpose of seeing what is in the interior of bodies. Light is a thing which is, so to say, attracted by every substance that gravitates (and we do not know anything that

[13] *Iceland* or *calc spar.* Native carbonate of lime in its primitive crystalline form.

does not). All matter affects light more or less by what we may consider as a kind of attraction, and I have arranged (FIG. 18) a very simple experiment upon the floor of the room for the purpose of illustrating this. I have put into

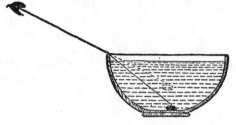

FIG. 18

that basin a few things which those who are in the body of the theatre will not be able to see, and I am going to make use of this power which matter possesses of attracting a ray of light. If Mr. Anderson pours some water, gently and steadily, into the basin, the water will attract the rays of light downward, and the piece of silver and the sealing-wax will appear to rise up into the sight of those who were before not high enough to see over the side of the basin to its bottom. [Mr. Anderson here poured water into the basin, and upon the lecturer asking whether any body could see the silver and sealing-wax, he was answered by a general affirmative.] Now I suppose that every body can see that they are not at all disturbed, while from the way they appear to have risen up you would imagine the bottom of the basin and the articles in it were two inches thick, although they are only one of our small silver dishes and a piece of sealing-wax which I have put there. The light which now goes to you from that piece of silver was obstructed by the edge of the basin when there was no water there, and you were unable to see any thing of it; but when we poured in water the rays were attracted down by it over the edge of the basin, and you were thus enabled to see the articles at the bottom.

I have shown you this experiment first, so that you might understand how glass attracts light, and might then see how other substances like rock-salt and calcareous spar, **mica,** and other stones, would affect the light; and, if Dr. Tyndall will be good enough to let us use his light again, we will first of all show you how it may be bent by a piece of glass (FIG. 19). [The electric lamp was again lit, and the beam

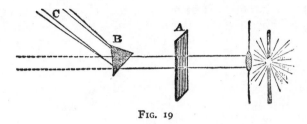

FIG. 19

of parallel rays of light which it emitted was bent about and decomposed by means of the prism.] Now, here you see, if I send the light through this piece of plain glass, A, it goes straight through without being bent (unless the glass be held obliquely, and then the phenomenon becomes more complicated); but if I take this piece of glass, B [a prism], you see it will show a very different effect. It no longer goes to that wall, but it is bent to this screen, C, and how much more beautiful it is now [throwing the prismatic spectrum on the screen]. This ray of light is bent out of its course by the attraction of the glass upon it; and you see I can turn and twist the rays to and fro in different parts of the room, just as I please. Now it goes there, now here. [The lecturer projected the prismatic spectrum about the theatre.] Here I have the rays once more bent on to the screen, and you see how wonderfully and beautifully that piece of glass not only bends the light by virtue of its attraction, but actually splits it up into different colors. Now I want you to understand that this piece of glass [the prism], being perfectly uniform in its internal structure, tells us about the action of these other bodies which are not uniform—which do not merely *cohere,* but also have within them, in different parts, different *degrees of cohesion,* and thus attract and bend the light with varying powers. We will now let the light pass through one or two of these things which I just now showed you broke so curiously; and, first of all, I will take a piece of mica. Here, you see, is our ray of light: we have first to make it what we call *polarized;* but about that you need not trouble yourselves; it is only to make our illustration more clear. Here, then, we have our polarized ray of light, and I can so adjust it as to make the screen upon which it is shining either light or dark, although I have nothing in the course of this ray of light but what is perfectly transparent [turning the *analyzer* round]. I will now make it so that it is quite dark, and we will, in the first instance, put a piece of common glass into the polarized ray so as to show you that it does not enable the light to get through. You see the screen remains dark. The glass, then, internally, has no effect upon light. [The glass was removed and a piece of mica introduced.] Now there is the mica which we split up so curiously into leaf after leaf, and see how that enables the light to pass through to the screen, and how, as Dr. Tyndall turns it round in his hand, you have those different colors, pink, and purple, and green, coming and going most beautifully; not that the mica is more transparent than the glass, but because of the different manner in which its particles are arranged by the force of cohesion.

Now we will see how calcareous spar acts upon this light —that stone which split up into rhombs, and of which you are each of you going to take a little piece home. [The mica was removed, and a piece of calc-spar introduced at A.] See how that turns the light round and round, and

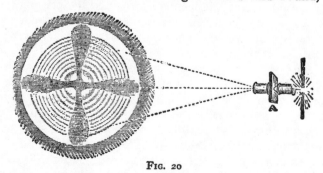

FIG. 20

produces these rings and that black cross (FIG. 20). Look at those colors: are they not most beautiful for you and for me? (for I enjoy these things as much as you do). In what

a wonderful manner they open out to us the internal arrangement of the particles of this calcareous spar by the force
of cohesion.

And now I will show you another experiment. Here is
that piece of glass which before had no action upon the light.
You shall see what it will do when we apply pressure to it.
Here, then, we have our ray of polarized light, and I will
first of all show you that the glass has no effect upon it
in its ordinary state; when I place it in the course of the
light, the screen still remains dark. Now Dr. Tyndall will
press that bit of glass between three little points, one point
against two, so as to bring a strain upon the parts, and you
will see what a curious effect that has. [Upon the screen
two white dots gradually appeared.] Ah! these points show
the position of the strain; in these parts the force of cohesion
is being exerted in a different degree to what it is in the
other parts, and hence it allows the light to pass through.
How beautiful that is! how it makes the light come through
some parts and leaves it dark in others, and all because
we weaken the force of cohesion between particle and particle. Whether you have this mechanical power of strain-
ing, or whether we take other means, we get the same
result; and, indeed, I will show you by another experiment
that if we heat the glass in one part, it will alter its internal
structure and produce a similar effect. Here is a piece of
common glass, and if I insert this in the path of
the polarized ray, I believe it will do nothing. There is
the common glass [introducing it]. No light passes
through; the screen remains quite dark; but I am going
to warm this glass in the lamp, and you know yourselves
that when you pour warm water upon glass you put a
strain upon it sufficient to break it sometimes—something
like there was in the case of the Prince Rupert's drops. [The
glass was warmed in the spirit lamp, and again placed across
the ray of light.] Now you see how beautifully the light goes
through those parts which are hot, making dark and light
lines just as the crystal did, and all because of the altera-
tion I have effected in its internal condition; for these dark
and light parts are a proof of the presence of forces acting
and dragging in different directions within the solid mass.

LECTURE III

COHESION—CHEMICAL AFFINITY

WE will first return for a few minutes to one of the experiments made yesterday. You remember what we put together on that occasion—powdered alum and warm water. Here is one of the basins then used. Nothing has been done to it since; but you will find, on examining it, that it no longer contains any powder, but a number of beautiful crystals. Here also are the pieces of coke which I put into the other basin; they have a fine mass of crystals about them. That other basin I will leave as it is. I will not pour the water from it, because it will show you that the particles of alum have done something more than merely crystallize together. They have pushed the dirty matter from them, laying it around the outside or outer edge of the lower crystals—squeezed out, as it were, by the strong attraction which the particles of alum have for each other.

And now for another experiment. We have already gained a knowledge of the manner in which the particles of bodies—of solid bodies—attract each other, and we have learned that it makes calcareous spar, and so forth, crystallize in these regular forms. Now let me gradually lead your minds to a knowledge of the means we possess of making this attraction alter a little in its force; either of increasing, or diminishing, or, apparently, of destroying it altogether. I will take this piece of iron [a rod of iron about two feet long and a quarter of an inch in diameter]. It has at present a great deal of strength, due to its attraction of cohesion; but if Mr. Anderson will make part of this red-hot in the fire, we shall then find that it will become soft, just as sealing-wax will when heated, and we shall also find that the more it is heated the softer it becomes. Ah!

36

but what does *soft* mean? Why, that the attraction between the particles is so weakened that it is no longer sufficient to resist the power we bring to bear upon it. [Mr. Anderson handed to the lecturer the iron rod, with one end red-hot, which he showed could be easily twisted about with a pair of pliers.] You see I now find no difficulty in bending this end about as I like, whereas I can not bend the cold part at all. And you know how the smith takes a piece of iron and heats it in order to render it soft for his purpose: he acts upon our principle of lessening the adhesion of the particles, although he is not exactly acquainted with the terms by which we express it.

And now we have another point to examine, and this water is again a very good substance to take as an illustration (as philosophers we call it all water, even though it be in the form of ice or steam). Why is this water hard? [pointing to a block of ice]; because the attraction of the particles to each other is sufficient to make them retain their places in opposition to force applied to it. But what happens when we make the ice warm? Why, in that case we diminish to such a large extent the power of attraction that the solid substance is destroyed altogether. Let me illustrate this: I will take a red-hot ball of iron [Mr. Anderson, by means of a pair of tongs, handed to the lecturer a red-hot ball of iron, about two inches in diameter], because it will serve as a convenient source of heat [placing the red-hot iron in the centre of the block of ice]. You see I am now melting the ice where the iron touches it. You see the iron sinking into it; and while part of the solid water is becoming liquid, the heat of the ball is rapidly going off. A certain part of the water is actually rising in steam, the attraction of some of the particles is so much diminished that they can not even hold together in the liquid form, but escape as vapor. At the same time, you see I can not melt all this ice by the heat contained in this ball. In the course of a very short time I shall find it will have become quite cold.

Here is the water which we have produced by destroying some of the attraction which existed between the particles of the ice, for below a certain temperature the particles of water increase in their mutual attraction and become ice; and above a certain temperature the attraction decreases and the water becomes steam. And exactly the same thing happens with platinum, and nearly every substance in nature; if the temperature is increased to a certain point it becomes liquid and a farther increase converts it into a gas. Is it not a glorious thing for us to look at the sea, the rivers, and so forth, and to know that this same body in the northern regions is all solid ice and icebergs, while here, in a warmer climate, it has its attraction of cohesion so much diminished as to be liquid water? Well, in diminishing this force of attraction between the particles of ice, we made use of another force, namely that of *heat;* and I want you now to understand that this force of heat is always concerned when water passes from the solid to the liquid state. If I melt ice in *other* ways I can not do without heat (for we have the means of making ice liquid without heat—that is to say, without using heat as a *direct* cause). Suppose, for illustration, I make a vessel out of this piece of tinfoil [bending the foil up into the shape of a dish]. I am making it metallic, because I want the heat which I am about to deal with to pass readily through it; and I am going to pour a little water on this board, and then place the tin vessel on it. Now if I put some of this ice into the metal dish, and then proceed to make it liquid by any of the various means we have at our command, it still must take the necessary quantity of heat from something, and in this case it will take the heat from the tray, and from the water underneath, and from the other things round about. Well, a little salt added to the ice has the power of causing it to melt, and we shall very shortly see the mixture become quite fluid, and you will then find that the water beneath will be frozen—frozen because it has been forced to give up that heat which is necessary to keep it in the liquid state to the ice on becoming liquid. I remember once, when I was a boy, hearing of a trick in a country ale-house: the point was how to melt ice in a quart pot by the fire and freeze it to the stool. Well, the way they did it was this: they put some pounded ice in a pewter pot, and added some salt to it, and the consequence was that when the salt was mixed with it, the ice in the pot melted (they did not tell me any thing about the salt and they set the pot by the fire, just to make the result more mysterious), and in a short time the pot and the stool were frozen together, as we shall very shortly find it to be the case here, and all because salt has the power of lessening the attraction between the particles of ice. Here you see the tin dish is frozen to the board; I can even lift the little stool up by it.

This experiment can not, I think, fail to impress upon your minds the fact that whenever a solid body loses some of that force of attraction by means of which it remains solid, heat is absorbed; and if on the other hand we convert a liquid into a solid, *e. g.,* water into ice, a corresponding amount of heat is given out.   I have an experiment showing this to be the case.   Here (**FIG. 21**) is a bulb, A, filled with

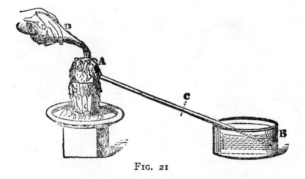

FIG. 21

air, the tube from which dips into some colored liquid in the vessel B.   And I dare say you know that if I put my hand on the bulb A, and warm it, the colored liquid which is now standing in the tube at C will travel forward.   Now we have discovered a means, by great care and research into the properties of various bodies, of preparing a solution of a salt([14]) which, if shaken or disturbed, will at once become a solid; and as I explained to you just now (for what is true of water is true of every other liquid), by reason of its becoming solid heat is evolved, and I can make this evident

[14] *Solution of a salt.* Acetate of soda. A solution saturated, or nearly so, at the boiling point, is necessary, and it must be allowed to cool, and remain at rest until the experiment is made.

to you by pouring it over this bulb; there! it is becoming
solid; and look at the colored liquid, how it is being driven
down the tube, and how it is bubbling out through the water
at the end; and so we learn this beautiful law of our
philosophy, that whenever we diminish the attraction of
cohesion we absorb heat, and whenever we increase that
attraction heat is evolved. This, then, is a great step in
advance, for you have learned a great deal in addition to
the mere circumstance that particles attract each other. But
you must not now suppose that because they are liquid they
have lost their attraction of cohesion; for here is the fluid
mercury, and if I pour it from one vessel into another, I find
that it will form a stream from the bottle down to the
glass—a continuous rod of fluid mercury, the particles of
which have attraction sufficient to make them hold together
all the way through the air down to the glass itself; and if
I pour water quietly from a jug, I can cause it to run in
a continuous stream in the same manner.   Again: let me
put a little water on this piece of plate glass, and then
take another plate of glass and put it on the water,
there! the upper plate is quite free to move, gliding about
on the lower one from side to side; and yet, if I take
hold of the upper plate and lift it up straight, the cohesion
is so great that the lower one is held up by it. See how
it runs about as I move the upper one, and this is all owing
to the strong attraction of the particles of the water. Let
me show you another experiment. If I take a little soap
and water—not that the soap makes the particles of the
water more adhesive one for the other, but it certainly has
the power of continuing in a better manner the attraction
of the particles (and let me advise you, when about to
experiment with soap-bubbles, to take care to have every
thing clean and soapy). I will now blow a bubble, and that
I may be able to talk and blow a bubble too, I will take a plate
with a little of the soapsuds in it, and will just soap the
edges of the pipe and blow a bubble on to the plate. Now
there is our bubble. Why does it hold together in this
manner? Why, because the water of which it is composed
has an attraction of particle for particle—so great, indeed,
that it gives to this bubble the very power of an India-rubber

ball; for you see, if I introduce one end of this glass tube into the bubble, that it has the power of contracting so powerfully as to force enough air through the tube to blow out a light (FIG. 22); the light is blown out. And look! see how the bubble is disappearing—see how it is getting smaller and smaller.

There are twenty other experiments I might show you to illustrate this power of cohesion of the particles of liquids. For instance, what would you propose to me if, having lost the stopper out of this alcohol bottle, I should want to close it speedily with something near at hand. Well, a bit of paper would not do, but a piece of linen cloth would, or some of this cotton wool which I have here. I will put a tuft of it into the neck of the alcohol bottle, and you see,

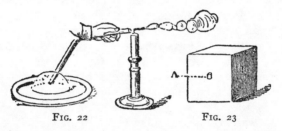

FIG. 22   FIG. 23

when I turn it upside down, that it is perfectly well stoppered so far as the alcohol is concerned; the air can pass through, but the alcohol can not. And if I were to take an oil vessel this plan would do equally well, for in former times they used to send us oil from Italy in flasks stoppered only with cotton wool (at the present time the cotton is put in after the oil has arrived here, but formerly it used to be sent so stoppered). Now if it were not for the particles of liquid cohering together, this alcohol would run out; and if I had time I could have shown you a vessel with the top, bottom, and sides altogether formed like a sieve, and yet it would hold water, owing to the cohesion.

You have now seen that the solid water can become fluid by the addition of heat, owing to this lessening the attractive force between its particles, and yet you see that there is a good deal of attractive force remaining behind. I want now to take you another step beyond. We saw that if we continued applying heat to the water (as indeed happened with our piece of ice here), that we did at last break up that attraction which holds the liquid together, and I am about to take some other (any other liquid would do, but ether makes a better experiment for my purpose) in order to illustrate what will happen when this cohesion is broken up. Now this liquid ether, if exposed to a very low temperature, will become a solid; but if we apply heat to it, it becomes vapor; and I want to show you the enormous bulk of the substance in this new form: when we make ice into water, we lessen its bulk; but when we convert water into steam, we increase it to an enormous extent. You see it is very clear that as I apply heat to the liquid I diminish its attraction of cohesion; it is now boiling, and I will set fire to the vapor, so that you may be enabled to judge of the space occupied by the ether in this form by the size of its flame; and you now see what an enormously bulky flame I get from that small volume of ether below. The heat from the spirit lamp is now being consumed, not in making the ether any warmer, but in converting it into vapor; and if I desired to catch this vapor and condense it (as I could without much difficulty), I should have to do the same as if I wished to convert steam into water and water into ice: in either case it would be necessary to increase the attraction of the particles by cold or otherwise. So largely is the bulk occupied by the particles increased by giving them this diminished attraction, that if I were to take a portion of water a cubic inch in bulk (A, FIG. 23), I should produce a volume of steam of that size, B [1,700 cubic inches; nearly a cubic foot], so greatly is the attraction of cohesion diminished by heat; and yet it still remains water. You can easily imagine the consequences which are due to this change in volume by heat— the mighty powers of steam and the tremendous explosions which are sometimes produced by this force of water. I want you now to see another experiment, which will perhaps give you a better illustration of the bulk occupied by a body when in the state of vapor. Here is a substance which we call iodine, and I am about to submit this solid body to the same kind of condition as regards heat that I did the water and the other [putting a few grains of iodine into a hot glass globe, which immediately became filled with the violet vapor], and you see the same kind of change produced. Moreover, it gives us the opportunity of observing how beautiful is the violet-colored vapor from this black substance, or rather the mixture of the vapor with air (for I would not wish you to understand that this globe is entirely filled with the vapor of iodine).

If I had taken mercury and converted it into vapor (as I could easily do), I should have a perfectly colorless vapor; for you must understand this about vapors, that bodies in what we call the vaporous or the gaseous state are always perfectly transparent, never cloudy or smoky; they are, however, often colored, and we can frequently have colored vapors or gases produced by colorless particles themselves mixing together, as in this case [the lecturer here inverted a glass cylinder full of binoxide of nitrogen ([15]) over a cylinder of oxygen, when the dark red vapor of hyponitrous acid was produced]. Here also you see a very excellent illustration of the effect of a power of nature which we have not as yet come to, but which stands next on our list— CHEMICAL AFFINITY. And thus you see we can have a violet vapor or an orange vapor, and different other kinds of vapor, but they are always perfectly transparent, or else they would cease to be vapors.

I am now going to lead you a step beyond this consideration of the attraction of the particles for each other. You see we have come to understand that, if we take water as an illustration, whether it be ice, or water, or steam, it is always to be considered by us as water. Well, now prepare your minds to go a little deeper into the subject. We have means of searching into the constitution of water beyond any that are afforded us by the action of heat, and among these one of the most important is that force which we call voltaic electricity, which we used at our last meeting for the purpose of obtaining light, and which we carried

[15] *Binoxide of nitrogen* and *hyponitrous acid*. Binoxide of nitrogen is formed when nitric acid and a little water are added to some copper turnings. It produces deep red fumes as soon as it comes in contact with the air, by combining with the oxygen of the latter to form hyponitrous acid. *Binoxide of nitrogen* is composed of two parts of oxygen and one part of nitrogen; *hyponitrous acid* is composed of one part of nitrogen and three parts of oxygen.

about the room by means of these wires. This force is produced by the battery behind me, to which, however, I will not now refer more particularly; before we have done we shall know more about this battery, but it must grow up in our knowledge as we proceed. Now here (FIG. 24) is a portion of water in this little vessel, C, and besides the water there are two plates of the metal platinum, which are connected with the wires (A and B) coming outside,

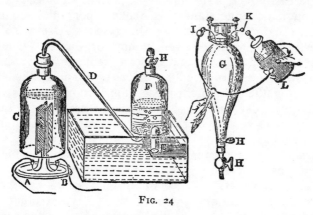

FIG. 24

and I want to examine that water, and the state and the condition in which its particles are arranged. If I were to apply heat to it you know what we should get; it would assume the state of vapor, but it would nevertheless remain water, and would return to the liquid state as soon as the heat was removed. Now by means of these wires (which are connected with the battery behind me, and come under the floor and up through the table) we shall have a certain amount of this new power at our disposal. Here you see it is [causing the ends of the wires to touch]—that is the electric light we used yesterday, and by means of these wires we can cause water to submit itself to this power; for the moment I put them into metallic connection (at A and B) you see the water boiling in that little vessel (C), and you hear the bubbling of the gas that is going through the tube (D). See how I am converting the water into vapor; and if I take a little vessel (E), and fill it with water, and put it into the trough over the end of the tube (D), there goes the vapor ascending into the vessel. And yet that is not steam, for you know that if steam is brought near cold water, it would at once condense, and return back again to water; this, then, can not be steam, for it is bubbling through the cold water in this trough; but it is a vaporous substance, and we must therefore examine it carefully, to see in what way the water has been changed. And now, in order to give you a proof that it is not steam, I am going to show you that it is combustible; for if I take this small vessel to a light, the vapor inside explodes in a manner that steam could never do.

I will now fill this large bell-jar (F) with water; and I propose letting the gas ascend into it, and I will then show you that we can reproduce the water back again from the vapor or air that is there. Here is a strong glass vessel (G), and into it we will let the gas (from F) pass. We will there fire it by the electric spark, and then, after the explosion, you will find that we have got the water back again; it will not be much, however, for you will recollect that I showed you how small a portion of water produced a very large volume of vapor. Mr. Anderson will now pump all the air out of this vessel (G), and when I have screwed it on to the top of our jar of gas (F), you will see, upon opening the stop-cocks (H H H), the water will jump up, showing that some of the gas has passed into the glass vessel. I will now shut these stop-cocks, and we shall be able to send the electric spark through the gas by means of the wires (I, K) in the upper part of the vessel, and you will see it burn with a most intense flash. [Mr. Anderson here brought a Leyden jar, which he discharged through the confined gas by means of the wires (I, K).] You saw the flash, and now that you may see that there is no longer any gas remaining, if I place it over the jar and open the stop-cocks again, up will go the gas, and we can have a second combustion; and so I might go on again and again, and I should continue t› accumulate more and more of the water to which the gas has returned. Now is not this curious? In this vessel (C) we can go on making from water a large bulk of *permanent gas,* as we call it, and then we can reconvert it into water in this way. [Mr.

Anderson brought in another Leyden jar, which, however, from some cause, would not ignite the gas. It was therefore recharged, when the explosion took place in the desired manner.] How beautifully we get our results when we are right in our proceedings! It is not that Nature is wrong when we make a mistake. Now I will lay this vessel (G) down by my right hand, and you can examine it by-and-by; there is not very much water flowing down, but there is quite sufficient for you to see.

Another wonderful thing about this mode of changing the condition of the water is this: that we are able to get the separate parts of which it is composed at a distance the one from the other, and to examine them, and see what they are like, and how many of them there are; and for this purpose I have here some more water in a slightly different apparatus to the former one (Fig. 25)

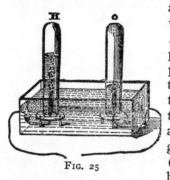

Fig. 25

and if I place this in connection with the wires of the battery (at A, B), I shall get a similar decomposition of the water at the two platinum plates. Now I will put this little tube (O) over there, and that will collect the gas together that comes from this side (A), and this tube (H) will collect the gas that comes from the other side (B), and I think we shall soon be able to see a difference. In this apparatus the wires are a good way apart from each other, and it now seems that *each* of them is capable of drawing off particles from the water and sending them off, and you see that one set of particles (H) is coming off twice as fast as those collected in the other tube (O). Something is coming out of the water *there* (at H) which burns [setting fire to the gas]; but what comes out of the water *here* (at O), although it will not burn, will support combustion very vigorously. [The lecturer here placed a match with a glowing tip in the gas, when it immediately rekindled.]

Here, then, we have two things, neither of them being water alone, but which we get out of the water. Water is therefore composed of two substances different to itself, which appear at separate places when it is made to submit to the force which I have in these wires; and if I take an inverted tube of water and collect this gas (H), you will see that it is by no means the same as the one we collected in the former apparatus (Fig. 24). That exploded with a loud noise when it was lighted, but this will burn quite noiselessly: it is called *hydrogen;* and the other we call *oxygen*—that gas which so beautifully brightens up all combustion, but does not burn of itself. So now we see that water consists of two kinds of particles attracting each other in a very different manner to the attraction of gravitation or cohesion, and this new attraction we call *chemical affinity,* or the force of chemical action between different bodies; we are now no longer concerned with the attraction of iron for iron, water for water, wood for wood, or like bodies for each other, as we were when dealing with the force of cohesion; we are dealing with another kind of attraction—the attraction between particles of a *different* nature one to the other. Chemical affinity depends entirely upon the energy with which particles of *different* kinds attract each other. Oxygen and hydrogen are particles of different kinds, and it is their attraction to each other which makes them chemically combine and produce water.

I must now show you a little more at large what chemical affinity is. I can prepare these gases from other substances as well as from water; and we will now prepare some oxygen: here is another substance which contains oxygen—chlorate of potash; I will put some of it into this glass retort, and Mr. Anderson will apply heat to it: we have here different jars filled with water, and when, by the application of heat, the chlorate of potash is decomposed, we will displace the water, and fill the jars with gas.

Now, when water is opened out in this way by means of the battery, which adds nothing to it materially, which takes nothing from it materially (I mean no *matter;* I am not speaking of *force*), which adds no *matter* to the water, it is changed in this way—the gas which you saw burning a little while ago, called *hydrogen,* is evolved in large quantity, and the other gas, *oxygen,* is evolved in only half the quantity; so that these two areas represent water, and these are always the proportions between the two gases.

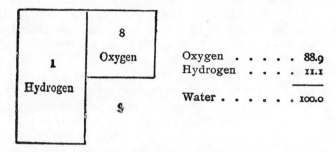

Oxygen	88.9
Hydrogen	11.1
Water	100.0

But oxygen is sixteen times the weight of the other—eight times as heavy as the particles of hydrogen in the water; and you therefore know that water is composed of nine parts by weight—one of hydrogen and eight of oxygen; thus:

Hydrogen.......... 46.2 cubic inches.......... = 1 grain
Oxygen............ 23.1   "   "   .... = 8 grains

Water (*steam*)..... 69.3   "   "   .... = 9 grains

Now Mr. Anderson has prepared some oxygen, and we will proceed to examine what is the character of this gas. First of all, you remember I told you that it does not burn, but that it affects the burning of other bodies. I will just set fire to the point of this little bit of wood, and then plunge it into the jar of oxygen, and you will see what this gas does in increasing the brilliancy of the combustion. It does not burn, it does not take fire, as the hydrogen would; but how vividly the combustion of the match goes on! Again, if I were to take this wax taper and light it, and turn it upside down in the air, it would in all probability put itself out, owing to the wax running down into the wick. [The lecturer here turned the lighted taper upside down, when in a few seconds it went out.] Now that will not happen in oxygen gas; you will see how differently it acts (FIG. 26). [The taper was again lighted, turned upside down, and then introduced into a jar of oxygen.] Look at that! See how the very wax itself burns, and falls down in a dazzling stream of fire, so powerfully does the oxygen support combustion. Again, here is another experiment which will serve to illustrate the force, if I may so call it, of oxygen. I have here a circular flame of spirit of wine, and with it I am about to show you the way in which iron burns, because it will serve very well as a comparison between the effect produced by air and oxygen. If I take this ring flame, I can shake, by means of a sieve, the fine *particles* of iron filings through it, and you will see the way in which they burn. [The lecturer here shook through the flame some iron filings, which took fire and fell

FIG. 26

through with beautiful scintillations.] But if I now hold the flame over a jar of oxygen [the experiment was repeated over a jar of oxygen, when the combustion of the filings as they fell into the oxygen became almost insupportably brilliant], you see how wonderfully different the effect is in the jar, because there we have oxygen instead of common air.

LECTURE IV

CHEMICAL AFFINITY—HEAT

WE shall have to pay a little more attention to the forces existing in water before we can have a clear idea on the subject. Besides the attraction which there is between its particles to make it hold together as a liquid or a solid, there is also another force, different from the former—one which, yesterday, by means of the voltaic battery, we overcame, drawing from the water two different substances, which, when heated by means of the electric spark, attracted each other, and rushed into combination to reproduce water. Now I propose to-day to continue this subject, and trace the various phenomena of chemical affinity; and for this purpose, as we yesterday considered the character of oxygen, of which I have here two jars (oxygen being those particles derived from the water which enable other bodies to burn), we will now consider the other constituent of water, and, without embarrassing you too much with the way in which these things are made, I will proceed now to show you our common way of making *hydrogen*. (I called it hydrogen yesterday: it is so called because it helps to generate water.)* I put into this retort some zinc, water and oil of vitriol, and immediately an action takes place, which produces an abundant evolution of gas, now coming over into this jar, and bubbling up in appearance exactly like the oxygen we obtained yesterday.

The processes, you see, are very different, though the result is the same, in so far as it gives us certain gaseous particles. Here, then, is the hydrogen. I showed you yesterday certain qualities of this gas; now let me exhibit you some other properties. Unlike oxygen, which is a supporter of combustion and will not burn, hydrogen itself is com-

_____
* ὑδωρ, "water," and γεννάω, "I generate."

50

bustible. There is a jar full of it; and if I carry it along in this manner and put a light to it, I think you will see it take fire—not with a bright light; you will, at all events,

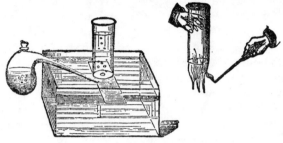

FIG. 27

hear it if you do not see it. Now that is a body entirely different from oxygen; it is extremely light; for, although yesterday you saw twice as much of this hydrogen produced on the one side as on the other by the voltaic battery, it was only one-eighth the weight of the oxygen. I carry this jar upside down. Why? Because I know that it is a very light body, and that it will continue in this jar upside down quite as effectually as the water will in that jar which is not upside down; and just as I can pour water from one vessel into another in the right position to receive it, so can I pour this gas from one jar into another when they are upside down. See what I am about to do. There is no hydrogen in this jar at present, but I will gently turn this jar of hydrogen up under this other jar (FIG. 28), and then we will examine the two. We shall see, on applying a light, that the hydrogen has left the jar in which it was at first, and has poured upward into the other, and there we shall find it.

FIG. 28

You now understand that we can have particles of very different kinds, and that they can have different bulks and weights; and there are two or three very interesting experiments which serve to illustrate this. For instance, if I blow soap bubbles with the breath from my mouth, you will see them fall, because I fill them

with common air, and the water which forms the bubble car-
ries it down. But now, if I inhale hydrogen gas into my lungs
(it does no harm to the lungs, although it does no good to
them), see what happens. [The lecturer inhaled some hydrogen, and, after one or two ineffectual attempts, succeeded in
blowing a splendid bubble, which rose majestically and slowly
to the ceiling of the theatre, where it burst.] That shows you
very well how light a substance this is; for, notwithstanding
all the heavy bad air from my lungs, and the weight of the
bubble, you saw how it was carried up. I want you now to
consider this phenomenon of weight as indicating how exceed-
ingly different particles are one from the other; and I will
take as illustrations these very common things, air, water, the
heaviest body, platinum, and this gas, and observe how they
differ in this respect; for if I take a piece of platinum of
that size (FIG. 29), it is equal to the weight of portions of

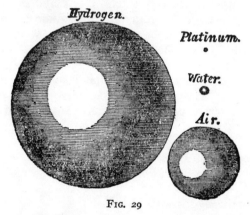

FIG. 29

water, air, and hydrogen of the bulks I have represented in
these spheres; and this illustration gives you a very good
idea of the extraordinary difference with regard to the
gravity of the articles having this enormous difference in
bulk. [The following tabular statement having reference to
this illustration appeared on the diagram-board.]

Hydrogen	. . . .	1		
Air	. . . . . .	14.4	1	
Water	. . . . .	11943	829	1
Platinum	. . . .	256774	17831	21.5

Whenever oxygen and hydrogen unite together they produce water, and you have seen the extraordinary difference between the bulk and appearance of the water so produced and the particles of which it consists chemically. Now we have never yet been able to reduce either oxygen or hydrogen to the liquid state; and yet their first impulse, when chemically combined, is to take up first this liquid condition and then the solid condition. We never combine these different particles together without producing water; and it is curious to think how often you must have made the experiment of combining oxygen and hydrogen to form water without knowing it. Take a candle, for instance, and a clean silver spoon (or a piece of clean tin will do), and, if you hold it over the flame, you immediately cover it with dew—not a smoke—which presently evaporates. This, perhaps, will serve to show it better. Mr. Anderson will put a candle under that jar, and you will see how soon the water is produced (FIG. 30). Look at that dimness on the sides of the glass, which will soon produce drops, and trickle down into the plate. Well, that dimness and these drops are *water,* formed by the union of the oxygen of the air with the hydrogen existing in the wax of which that candle is formed.

FIG. 30

And now, having brought you, in the first place, to the consideration of chemical attraction, I must enlarge your ideas so as to include all substances which have this attraction for each other; for it changes the character of bodies, and alters them in this way and that way in the most extraordinary manner, and produces other phenomena wonderful to think about. Here is some chlorate of potash, and there some sulphuret of antimony([16]). We will mix these two different sets of particles together and I want to show you, in a general sort of way, some of the phenomena which

[16] *Chlorate of potash and sulphuret of antimony.* Great care must be taken in mixing these substances, as the mixture is dangerously explosive. They must be powdered separately and mixed together with a feather on a sheet of paper, or by passing them several times through a small sieve.

take place when we make different particles act together.
Now I can make these bodies act upon each other in several
ways. In this case I am going to apply heat to the mixture;
but if I were to give a blow with a hammer, the same result
would follow. [A lighted match was brought to the mixture,
which immediately exploded with a sudden flash, evolving a
dense white smoke.] There you see the result of the action
of chemical affinity overcoming the attraction of cohesion of
the particles. Again, here is a little sugar([17]), quite a
different substance from the black sulphuret of antimony,
and you shall see what takes place when we put the two together. [The mixture was touched with sulphuric acid,
when it took fire, and burnt gradually and with a brighter
flame than in the former instance.] Observe this chemical
affinity traveling about the mass, and setting it on fire, and
throwing it into such wonderful agitation!

I must now come to a few circumstances which require
careful consideration. We have already examined one of
the effects of this chemical affinity, but, to make the matter
more clear, we must point out some others. And here are
two salts dissolved in water([18]). They are both colorless
solutions, and in these glasses you can not see any difference
between them. But if I mix them, I shall have chemical
attraction take place. I will pour the two together into this
glass, and you will at once see, I have no doubt, a certain
amount of change. Look, they are already becoming milky,
but they are sluggish in their action—not quick as the others
were—for we have endless varieties of rapidity in chemical
action. Now, if I mix them together, and stir them so as
to bring them properly together, you will soon see what a
different result is produced. As I mix them they get thicker
and thicker, and you see the liquid is hardening and stiff-
ening, and before long I shall have it quite hard; and before the end of the lecture it will be a solid stone—a wet
stone, no doubt, but more or less solid—in consequence
of the chemical affinity. Is not this changing two liquids

[17] The mixture of chlorate of potash and sugar does not require the same
precautions. They may be rubbed together in a pestle and mortar without
fear. One part of chlorate of potash and three parts of sugar will answer.
The mixture need only be touched with a glass rod dipped in oil of vitriol.
[18] *Two salts dissolved in water.* Sulphate of soda and chloride of calcium.
The solutions must be saturated for the experiment to succeed well.

into a solid body a wonderful manifestation of chemical affinity?

There is another remarkable circumstance in chemical affinity, which is, that it is capable of either waiting or acting at once. And this is very singular, because we know of nothing of the kind in the forces either of gravitation or cohesion. For instance: here are some oxygen particles, and here is a lump of carbon particles. I am going to put the carbon particles into the oxygen; they *can* act, but they *do* not—they are just like this unlighted candle. It stands here quietly on the table, waiting until we want to light it. But it is not so in this other case: here is a substance, gaseous like the oxygen, and if I put these particles of metal into it the two combine at once. The copper and the chlorine unite by their power of chemical affinity, and produce a body entirely unlike either of the substances used. And in this other case, it is not that there is any deficiency of affinity between the carbon and oxygen, for the moment I choose to put them in a condition to exert their affinity, you will see the difference. [The piece of charcoal was ignited, and introduced into the jar of oxygen, when the combustion proceeded with vivid scintillations.]

Now this chemical action is set going exactly as it would be if I had lighted the candle, or as it is when the servant puts coals on and lights the fire: the substances wait until we do something which is able to start the action. Can any thing be more beautiful than this combustion of charcoal in oxygen? You must understand that each of these little sparks is a portion of the charcoal, or the bark of the charcoal thrown off white hot into the oxygen, and burning in it most brilliantly, as you see. And now let me tell you another thing, or you will go away with a very imperfect notion of the powers and effects of this affinity. There you see some charcoal burning in oxygen. Well, a piece of lead will burn in oxygen just as well as the charcoal does, or indeed better, for absolutely that piece of lead will act at once upon the oxygen as the copper did in the other vessel with regard to the chlorine. And here, also, a piece of iron—if I light it and put it into the oxygen, it will burn away just as the carbon did. And I will take some lead, and show you that it will burn in the common atmospheric oxygen at the ordinary temperature. These are the lumps of lead which you remember we had the other day—the two pieces which clung together. Now these pieces, if I take them to-day and press them together, will not stick, and the reason is that they have attracted from the atmosphere a part of the oxygen there present, and have become coated as with a varnish by the oxide of lead, which is formed on the surface by a real process of combustion or combination. There you see the iron burning very well in oxygen, and I will tell you the reason why those scissors and that lead do not take fire while they are lying on the table. Here the lead is in a lump, and the coating of oxide remains on its surface, while there you see the melted oxide is clearing itself off from the iron, and allowing more and more to go on burning. In this case, however, [holding up a small glass tube containing lead pyrophorus([19])], the lead has been very carefully produced in fine powder, and put into a glass tube, and hermetically sealed so as to preserve it, and I expect you will see it take fire at once. This has been made about a month ago, and has thus had time to sink down to its normal temperature; what you see, therefore, is the result of chemical affinity alone. [The tube was broken at the end, and the lead poured out on to a piece of paper, whereupon it immediately took fire.] Look! look at the lead burning! Why, it has set fire to the paper! Now that is nothing more than the common affinity always existing between very clean lead and the atmospheric oxygen; and the reason why this iron does not burn until it is made red hot is because it has got a coating of oxide about it, which stops the action of the oxygen—putting a varnish, as it were, upon its surface, as we varnish a picture—absolutely forming a substance which prevents the natural chemical affinity between the bodies from acting.

I must now take you a little farther in this kind of illustration, or consideration I would rather call it, of chemical affinity. This attraction between different particles exists also most curiously in cases where they are previously com-

[19] *Lead pyrophorus.* This is tartrate of lead which has been heated in a glass tube to dull redness as long as vapors are emitted. As soon as they cease to be evolved the end of the tube is sealed, and it is allowed to cool.

bined with other substances. Here is a little chlorate of potash containing the oxygen which we found yesterday could be procured from it; it contains the oxygen there combined and held down by its chemical affinity with other things, but still it can combine with sugar, as you saw. This affinity can thus act *across* substances, and I want you to see how curiously what we call combustion acts with respect to this force of chemical affinity. If I take a piece of phosphorus and set fire to it, and then place a jar of air over the phosphorus, you see the combustion which we are having there on account of chemical affinity (combustion being in all cases the result of chemical affinity). The phosphorus is escaping in that vapor, which will condense into a snow-like mass at the close of the lecture. But suppose I limit the atmosphere, what then? why, even the phosphorus will go out. Here is a piece of camphor, which will burn very well in the atmosphere, and even on water it will float and burn, by reason of some of its particles gaining access to the air. But if I limit the quantity of air by placing a jar over it, as I am now doing, you will soon find the camphor will go out. Well, why does it go out? not for want of air, for there is plenty of air remaining in the jar. Perhaps you will be shrewd enough to say for want of oxygen.

This, therefore, leads us to the inquiry as to whether oxygen can do more than a certain amount of work. The oxygen there (FIG. 30) can not go on burning an unlimited quantity of candle, for that has gone out, as you see; and its amount of chemical attraction or affinity is just as strikingly limited: it can no more be fallen short of or exceeded than can the attraction of gravitation. You might as soon attempt to destroy gravitation, or weight, or all things that exist, as to destroy the exact amount of force exerted by this oxygen. And when I pointed out to you that eight by weight of oxygen to one by weight of hydrogen went to form water, I meant this, that neither of them would combine in different proportions with the other, for you can not get ten of hydrogen to combine with six of oxygen, or ten of oxygen to combine with six of hydrogen; it must be eight of oxygen and one of hydrogen. Now suppose I limit the action in this way: this piece of cotton wool burns, as you see, very well in the atmosphere; and I have known of cases of cotton-mills being fired as if with gunpowder through the very finely divided particles of cotton being diffused through the atmosphere in the mill, when it has sometimes happened that a flame has caught these raised particles, and it has run from one end of the mill to the other and blown it up. That, then, is on account of the affinity which the cotton has for the oxygen; but suppose I set fire to this piece of cotton which is rolled up tightly; it does not go on burning, because I have limited the supply of oxygen, and the inside is prevented from having access to the oxygen, just as it was in the case of the lead by the oxide. But here is some cotton which has been imbued with oxygen in a certain manner. I need not trouble you now with the way it is prepared; it is called gun-cotton([20]). See how that burns [setting fire to a piece]; it is very different from the other, because the oxygen which must be present in its proper amount is put there beforehand. And I have here some pieces of paper which are prepared like the gun-cotton([21]), and imbued with bodies containing oxygen. Here is some which has been soaked in nitrate of strontia: you will see the beautiful red color of its flame; and here is another which I think contains baryta, which gives that fine green light; and I have here some more which has been soaked in nitrate of copper: it does not burn quite so brightly, but still very beautifully. In all these cases the combustion goes on independent of the oxygen of the atmosphere. And here we have some gunpowder put into a case, in order to show that it is capable of burning under water. You know that we put it into a gun, shutting off the atmosphere with shot, and yet the oxygen which it contains supplies the particles with that without which chemical action could not proceed. Now I have a vessel of water here, and am going to make the experiment of putting this fuse under the water, and you

[20] *Gun-cotton* is made by immersing cotton-wool in a mixture of sulphuric acid and the strongest nitric acid or of sulphuric acid and nitrate of potash.

[21] *Paper prepared like gun-cotton.* It should be bibulous paper, and must be soaked for ten minutes in a mixture of ten parts, by measure, of oil of vitriol with five parts of strong fuming nitric acid. The paper must afterward be thoroughly washed with warm distilled water, and then carefully dried at a gentle heat. The paper is then saturated with chlorate of strontia, or chlorate of baryta, or nitrate of copper, by immersion in a warm solution of these salts. (See Chemical News, vol. i., p. 36.)

will see whether that water can extinguish it; here it is burning out of the water, and there it is burning under the water; and so it will continue until exhausted, and all by reason of the requisite amount of oxygen being contained within the substance. It is by this kind of attraction of the different particles one to the other that we are enabled to trace the laws of chemical affinity, and the wonderful variety of the exertions of these laws.

Now I want you to observe that one great exertion of this power which is known as *chemical affinity* is to produce HEAT and light; you know, as a matter of fact, no doubt, that when bodies burn they give out heat, but it is a curious thing that this heat does not continue; the heat goes away as soon as the action stops, and you see, thereby, that it depends upon the action *during the time* it is going on. It is not so with gravitation; this force is continuous, and is just as effective in making that lead press on the table as it was when it first fell there. Nothing occurs there which disappears when the action of falling is over; the pressure is upon the table, and will remain there until the lead is removed; whereas, in the action of chemical affinity to give light and heat, they go away immediately the action is over. This lamp *seems* to evolve heat and light continuously, but it is owing to a constant stream of air coming into it on all sides, and this work of producing light and heat by chemical affinity will subside as soon as the stream of air is interrupted. What, then, is this curious condition of heat? Why, it is the evolution of another power of matter—of a power new to us, and which we must consider as if it were now for the very first time brought under our notice. What is heat? We recognize heat by its power of liquefying solid bodies and vaporizing liquid bodies; by its power of setting in action, and very often overcoming, chemical affinity. Then how do we obtain heat? We obtain it in various ways; most abundantly by means of the chemical affinity we have just before been speaking about, but we can also obtain it in many other ways. Friction will produce heat. The Indians rub pieces of wood together until they make them hot enough to take fire; and such things have been known as two branches of a tree rubbing together so hard as to set the tree on fire. I do not suppose I shall set these two pieces of wood on fire by friction, but I can readily produce heat enough to ignite some phosphorus. [The lecturer here rubbed two pieces of cedar wood strongly against each other for a minute, and then placed on them a piece of phosphorus, which immediately took fire.] And if you take a smooth metal button stuck on a cork, and rub it on a piece of soft deal wood, you will make it so hot as to scorch wood and paper, and burn a match.

I am now going to show you that we can obtain heat, not by chemical affinity alone, but by the pressure of air. Suppose I take a pellet of cotton and moisten it with a little ether, and put it into a glass tube (FIG. 31), and then take a piston and press it down suddenly, I expect I shall be able to burn a little of that ether in the vessel. It wants a suddenness of pressure, or we shall not do what we require. [The piston was forcibly pressed down, when a flame, due to the combustion of the ether, was visible in the lower part of the syringe.] All we want is to get a little ether in vapor, and give fresh air each time, and so we may go on again and again, getting heat enough by the compression of air to fire the ether-vapor.

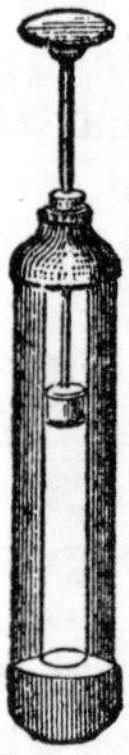

FIG. 31

This, then, I think will be sufficient, accompanied with all you have previously seen, to show you how we procure heat. And now for the effects of this power. We need not consider many of them on the present occasion, because, when you have seen its power of changing ice into water and water into steam, you have seen the two principal results of the application of heat. I want you now to see how it expands all bodies—all bodies but one, and that under limited circumstances. Mr. Anderson will hold a lamp under that retort, and you will see, the moment he does so, that the air will issue abundantly from the neck which is under water, because the heat which he applies to the air causes it to expand. And here is a brass rod (FIG. 32) which goes through that hole, and fits also accurately into this gauge; but if I make it warm with this spirit lamp, it will only go in the gauge or through the hole with difficulty; and if

I were to put it into boiling water it would not go through at all. Again, as soon as the heat escapes from bodies, they

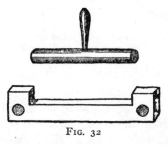

Fig. 32

collapse: see how the air is contracting in the vessel now that Mr. Anderson has taken away his lamp; the stem of it is filling with water. Notice too, now, that although I can not get the tube through this hole or into the gauge, the moment I cool it, by dipping it into water, it goes through with perfect facility, so that we have a perfect proof of this power of heat to contract and expand bodies.

LECTURE V

MAGNETISM—ELECTRICITY

I WONDER whether we shall be too deep to-day or not. Remember that we spoke of the attraction by gravitation of *all* bodies to all bodies by their simple approach. Remember that we spoke of the attraction of particles of the *same* kind to each other—that power which keeps them together in masses—iron attracted to iron, brass to brass, or water to water. Remember that we found, on looking into water, that there were particles of two different kinds attracted to each other; and this was a great step beyond the first simple attraction of gravitation, because here we deal with attraction between *different* kinds of matter. The hydrogen could attract the oxygen and reduce it to water, but it could not attract any of its own particles, so that there we obtained a first indication of the existence of *two* attractions.

To-day we come to a kind of attraction even more curious than the last, namely, the attraction which we find to be of a double nature—of a curious and dual nature. And I want, first of all, to make the nature of this doubleness clear to you. Bodies are sometimes endowed with a wonderful attraction, which is not found in them in their ordinary state. For instance, here is a piece of shellac, having the attraction of gravitation, having the attraction of cohesion, and if I set fire to it, it would have the attraction of chemical affinity to the oxygen in the atmosphere. Now all these powers we find *in* it as if they were parts of its substance; but there is another property which I will try and make evident by means of this ball, this bubble of air [a light India-rubber ball, inflated and suspended by a thread]. There is no attraction between this ball and this shellac at present; there may be a little wind in the rooms slightly moving the ball about, but

62

there is no attraction. But if I rub the shellac with a piece of flannel [rubbing the shellac, and then holding it near the ball], look at the attraction which has arisen out of the shellac simply by this friction, and which I may take away as easily by drawing it gently through my hand. [The lecturer repeated the experiment of exciting the shellac, and then removing the attractive power by drawing it through his hand.] Again, you will see I can repeat this experiment with another substance; for if I take a glass rod, and rub it with a piece of silk covered with what we call amalgam, look at the attraction which it has; how it draws the ball toward it; and then, as before, by quietly rubbing it through the hand, the attraction will be all removed again, to come back by friction with this silk.

But now we come to another fact. I will take this piece of shellac, and make it attractive by friction; and remember that, whenever we get an attraction of gravity, chemical affinity, adhesion, or electricity (as in this case), the body which attracts is attracted also, and just as much as that ball was attracted by the shellac, the shellac was attracted by the ball. Now I will suspend this piece of excited shellac in a little paper stirrup, in this way (FIG. 33), in order to

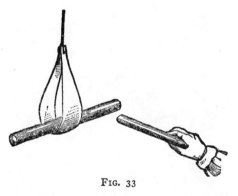

FIG. 33

make it move easily, and I will take another piece of shellac, and, after rubbing it with flannel, will bring them near together: you will think that they ought to attract each other; but now what happens? It does not attract; on the contrary, it very strongly *repels,* and I can thus drive it round to any extent. These, therefore, repel each other, although they are so strongly attractive—repel each other to the extent of driving this heavy piece of shellac round and round in this way. But if I excite this piece of shellac as before, and take this piece of glass and rub it with silk, and then bring them near, what think you will happen? [The lecturer held the excited glass near the excited shellac, when they attracted each other strongly.] You see, therefore, what a difference there is between these two attractions; they are actually two *kinds* of attraction concerned in this case, quite different to any thing we have met with before, but the *force* is the same. We have here, then, a double attraction—a dual attraction or force—one attracting and the other repelling.

Again, to show you another experiment which will help to make this clear to you: Suppose I set up this rough indicator again [the excited shellac suspended in the stirrup]: it is rough, but delicate enough for my purpose; and suppose I take this other piece of shellac, and take away the power, which I can do by drawing it gently through the hand; and suppose I take a piece of flannel (FIG. 34), which I have

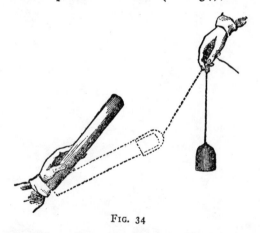

FIG. 34

shaped into a cap for it and made dry. I will put this shellac into the flannel, and here comes out a very beautiful result. I will rub this shellac and the flannel together (which I can do by twisting the shellac round), and leave them in contact;

and then if I ask, by bringing them near our indicator, what is the attractive force? it is nothing; but if I take them apart, and then ask what will they do when they are separated? why, the shellac is strongly repelled, as it was before, but the cap is strongly attractive; and yet, if I bring them both together again, there is no attraction; it has all disappeared [the experiment was repeated]. Those two bodies, therefore, still contain this attractive power; when they were parted, it was evident to your senses that they had it, though they do not attract when they are together.

This, then, is sufficient, in the outset, to give you an idea of the nature of the force which we call ELECTRICITY. There is no end to the things from which you can evolve this power. When you go home, take a stick of sealing-wax—I have rather a large stick, but a smaller one will do—and make an indicator of this sort (FIG. 35). Take a watch-glass (or your watch itself will do; you only want some-

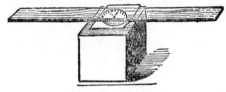

FIG. 35

thing which shall have a round face); and now, if you place a piece of flat glass upon that, you have a very easily moved centre; and if I take this lath and put it on the flat glass (you see I am searching for the centre of gravity of this lath; I want to balance it upon the watch-glass), it is very easily moved round; and if I take this piece of sealing-wax and rub it against my coat, and then try whether it is attractive [holding it near the lath], you see how strong the attraction is; I can even draw it about. Here, then, you have a very beautiful indicator, for I have, with a small piece of sealing-wax and my coat, pulled round a plank of that kind, so you need be in no want of indicators to discover the presence of this attraction. There is scarcely a substance which we may not use. Here are some indicators (FIG. 36). I bend round a strip of paper into a hoop, and we have as good an indicator as can be required. See how it rolls along, traveling after the sealing-wax! If I make them smaller, of course we have them running faster, and sometimes they are actually attracted up into the air. Here, also, is a little collodion balloon. It is so electrical that it will scarcely leave

FIG. 36

my hand unless to go to the other. See how curiously electrical it is; it is hardly possible for me to touch it without making it electrical; and here is a piece which clings to any thing it is brought near, and which it is not easy to lay down.

And here is another substance, gutta-percha, in thin strips: it is astonishing how, by rubbing this in your hands, you make it electrical; but our time forbids us to go farther into this subject at present; you see clearly there are two kinds of electricities which may be obtained by rubbing shellac with flannel or glass with silk.

Now there are some curious bodies in nature (of which I have two specimens on the table) which are called *magnets* or *loadstones;* ores of iron, of which there is a great deal sent from Sweden. They have the attraction of gravitation, and attraction of cohesion, and certain chemical attraction; but they also have a great attractive power, for this little key is held up by this stone. Now that is not chemical attraction; it is not the attraction of chemical affinity, or of aggregation of particles, or of cohesion, or of electricity (for it will not attract this ball if I bring it near it), but it is a separate and dual attraction, and, what is more, one which is not readily removed from the substance, for it has existed in it for ages and ages in the bowels of the earth. Now we can make artificial magnets (you will see me to-morrow make artificial magnets of extraordinary power). And let us take one of these artificial magnets and examine it, and see where the power is in the mass, and whether it is a dual power. You see it attracts these keys, two or three in succession, and it will attract a very large piece of iron. That, then, is a very different thing indeed to what you saw in the case of the shellac, for *that* only attracted a light ball, but here I have

several ounces of iron held up. And if we come to examine
this attraction a little more closely, we shall find it presents
some other remarkable differences; first of all, one end of
this bar (FIG. 37) attracts this key, but the middle does not
attract. It is not, then, the *whole* of the substance which
attracts. If I place this little key in the middle it does not
adhere; but if I place it *there,* a little nearer the end, it does,
though feebly. Is it not, then, very curious to find that
there is an attractive power at the extremities which is not
in the middle—to have thus in one bar two places in which
this force of attraction resides? If I take this bar and
balance it carefully on a point, so that it will be free to
move round, I can try what action this piece of iron has
on it. Well, it attracts one end, and it also attracts the
other end, just as you saw the shellac and the glass did, with
the exception of its not attracting in the middle. But if now,

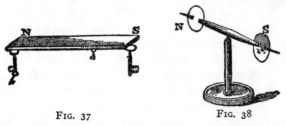

FIG. 37   FIG. 38

instead of a piece of iron, I take a *magnet,* and examine it
in a similar way, you see that one of its ends *repels* the
suspended magnet; the force, then, is no longer attraction,
but repulsion; but, if I take the other end of the magnet
and bring it near, it shows attraction again.

You will see this better, perhaps, by another kind of experiment. Here (FIG. 38) is a little magnet, and I have
colored the ends differently, so that you may distinguish one
from the other. Now this end (S) of the magnet (FIG. 37)
attracts the *uncolored* end of the little magnet. You see it
pulls toward it with great power; and, as I carry it round,
the uncolored end still follows. But now, if I gradually bring
the middle of the bar magnet opposite the uncolored end of
the needle, it has no effect upon it, either of attraction or
repulsion, until, as I come to the opposite extremity (N), you

see that it is the *colored* end of the needle which is pulled toward it. We are now, therefore, dealing with two kinds of power, attracting different ends of the magnet—a double power, already existing in these bodies, which takes up the form of attraction and repulsion. And now, when I put up this label with the word MAGNETISM, you will understand that it is to express this double power.

Now with this loadstone you may make magnets artificially. Here is an artificial magnet (FIG. 39) in which both ends have been brought together in order to increase the

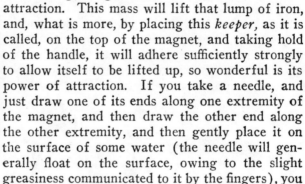

FIG. 39

attraction. This mass will lift that lump of iron, and, what is more, by placing this *keeper,* as it is called, on the top of the magnet, and taking hold of the handle, it will adhere sufficiently strongly to allow itself to be lifted up, so wonderful is its power of attraction. If you take a needle, and just draw one of its ends along one extremity of the magnet, and then draw the other end along the other extremity, and then gently place it on the surface of some water (the needle will generally float on the surface, owing to the slight greasiness communicated to it by the fingers), you will be able to get all the phenomena of attraction and repulsion by bringing another magnetized needle near to it.

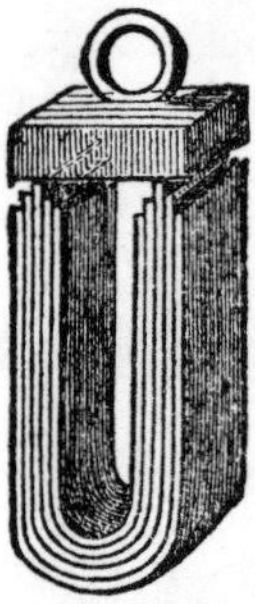

I want you now to observe that, although I have shown you in these magnets that this double power becomes evident principally at the extremities, yet the *whole* of the magnet is concerned in giving the power. That will at first seem rather strange; and I must therefore show you an experiment to prove that this is not an accidental matter, but that the whole of the mass is really concerned in this force, just as in falling the whole of the mass is acted upon by the force of gravitation. I have here (FIG. 40) a steel bar, and I am going to make it a magnet by rubbing it on the large magnet (FIG. 39). I have now made the two ends magnetic in opposite ways. I do not at present

FIG. 40

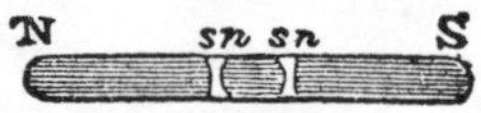

know one from the other, but we can soon find out. You see, when I bring it near our magnetic needle (FIG. 38), one end repels and the other attracts; and the middle will neither attract nor repel—it *can not,* because it is *half way between the two ends.* But now, if I break out that piece (*n, s*), and then examine it, see how strongly one end (*n*) pulls at this end (S, FIG. 38), and how it repels the other end (N). And so it can be shown that every part of the magnet contains this power of attraction and repulsion, but that the power is only rendered evident at the end of the mass. You will understand all this in a little while; but what you have now to consider is that every part of this steel is in itself a magnet. Here is a little fragment which I have broken out of the very centre of the bar, and you will still see that one end is attractive and the other is repulsive. Now is not this power a most wonderful thing? And very strange, the means of taking it from one substance and bringing it to other matters. I can not make a piece of iron or any thing else heavier or lighter than it is; its cohesive power it must and does have; but, as you have seen by these experiments, we can add or subtract this power of magnetism, and almost do as we like with it.

And now we will return for a short time to the subject treated of at the commencement of this lecture. You see

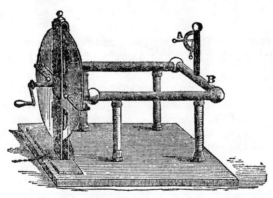

FIG. 41

here (FIG. 41) a large machine arranged for the purpose of rubbing glass with silk, and for obtaining the power called *electricity;* and the moment the handle of the machine is turned a certain amount of electricity is evolved, as you will see by the rise of the little straw indicator (at A). Now I know, from the appearance of repulsion of the pith ball at the end of the straw, that electricity is present in those brass conductors (BB), and I want you to see the manner in which that electricity can pass away [touching the conductor (B) with his finger, the lecturer drew a spark from it, and the straw electrometer immediately fell]. There, it has all gone; and that I have really taken it away you shall see by an experiment of this sort. If I hold this cylinder of brass by the glass handle, and touch the conductor with it, I take away a little of the electricity. You see the spark in which it passes, and observe that the pith-ball indicator has fallen a little, which seems to imply that so much electricity is lost; but it is not lost; it is here in this brass, and I can take it away and carry it about, not because it has any substance of its own, but by some strange property which we have not before met with as belonging to any other force. Let us see whether we have it here or not. [The lecturer brought the charged cylinder to a jet from which gas was issuing; the spark was seen to pass from the cylinder to the jet, but the gas did not light.] Ah! the gas did not light, but you saw the spark; there is, perhaps, some draught in the room which blew the gas on one side, or else it would light; we will try this experiment afterward. You see from the spark that I can transfer the power from the machine to this cylinder, and then carry it away and give it to some other body.

You know very well, as a matter of experiment, that we can transfer the power of heat from one thing to another; for if I put my hand near the fire it becomes hot. I can show you this by placing before us this ball, which has just been brought red-hot from the fire. If I press this wire to it some of the heat will be transferred from the ball, and I have only now to touch this piece of gun-cotton with the hot wire, and you see how I can transfer the heat from the ball to the wire, and from the wire to the cotton. So you see that some powers are transferable, and others are not. Observe how long the heat stops in this ball. I might touch it with the wire or with my finger, and if I did so quickly I should merely burn the surface of the skin; whereas, if I touch that cylinder, however rapidly, with my finger, the electricity is gone at once—dispersed on the instant, in a manner wonderful to think of.

I must now take up a little of your time in showing you the manner in which these powers are transferred from one thing to another; for the manner in which *force* may be conducted or transmitted is extraordinary, and most essential for us to understand. Let us see in what manner these powers travel from place to place. Both heat and electricity can be conducted; and here is an arrangement I have made to show how the former can travel. It consists of a bar of copper (FIG. 42); and if I take a spirit lamp (this is one way

FIG. 42

of obtaining the power of heat) and place it under that little chimney, the flame will strike against the bar of copper and keep it hot. Now you are aware that power is being transferred from the flame of that lamp to the copper, and you will see by-and-by that it is being conducted along the copper from particle to particle; for, inasmuch as I have fastened these wooden balls by a little wax at particular distances from the point where the copper is first heated, first one ball will fall and then the more distant ones, as the heat travels along, and thus you will learn that the heat travels gradually through the copper. You will see that this is a very slow conduction of power as compared with electricity. If I take cylinders of wood and metal, joined together at the ends, and wrap a piece of paper round, and then apply the heat of this lamp to the place where the metal and wood join, you will see how the heat will accumulate where the wood is, and burn the paper with which I have covered it; but where the metal is beneath, the heat is conducted away too fast for the paper to be burned. And so, if I take a piece of wood and a piece of metal joined together, and put it so that the flame shall play equally both upon one and the other, we shall soon find that the metal will become hot before the wood; for if I put a piece of phosphorus on the wood and another piece on the copper, you will find that the phosphorus on the copper will take fire before that on the wood is melted; and this shows you how badly the wood conducts heat. But with regard to the traveling of electricity from place to place, its rapidity is astonishing. I will, first of all, take these pieces of glass and metal, and you will soon understand how it is that the glass does not lose the power which it acquired when it is rubbed by the silk; by one or two experiments I will show you. If I take this piece of brass and bring it near the machine, you see how the electricity leaves the latter and passes to the brass cylinder. And again: if I take a rod of metal and touch the machine with it, I lower the indicator; but when I touch it with a rod of glass, no power is drawn away, showing you that the electricity is conducted by the glass and the metal in a manner entirely different; and, to make you see that more clearly, we will take one of our Leyden jars. Now I must not embarrass your minds with this subject too much, but if I take a piece of metal and bring it against the knob at the top and the metallic coating at the bottom, you will see the electricity passing through the air as a brilliant spark. It takes no sensible time to pass through this; and if I were to take a long metallic wire, no matter what the length, at least as far as we are concerned, and if I make one end of it touch the outside, and the other touch the knob at the top, see how the electricity passes! It has flashed instantaneously through the whole length of this wire. Is not this different from the transmission of heat through this copper bar (FIG. 42) which has taken a quarter of an hour or more to reach the first ball?

Here is another experiment for the purpose of showing the conductibility of this power through some bodies and not through others. Why do I have this arrangement made of brass? [pointing to the brass work of the electrical machine, FIG. 41.] Because it conducts electricity. And why do I have these columns made of glass? Because they obstruct the passage of electricity. And why do I put that paper tassel (FIG. 43) at the top of the pole, upon a glass rod, and connect it with this machine by means of a wire? You see at

once that as soon as the handle of the machine is turned, the electricity which is evolved travels along this wire and up the wooden rod, and goes to the tassel at the top, and you see the power of repulsion with which it has endowed these strips of paper, each spreading outward to the ceiling and sides of the room. The outside of that wire is covered with gutta-percha; it would not serve to keep the force from you when touching it with your hands, because it would burst through; but it answers our purpose for the present. And so you perceive how easily I can manage to send this power of electricity from place to place by choosing the materials which can conduct the power. Suppose I want to fire a portion of gunpowder, I can readily do it by this transferable power of electricity. I

FIG. 43

will take a Leyden jar, or any other arrangement which gives us this power, and arrange wires so that they may carry the power to the place I wish; and then placing a little gunpowder on the extremities of the wires, the moment I make the connection by this discharging rod I shall fire the gunpowder [the connection was made and the gunpowder ignited]. And if I were to show you a stool like this, and were to explain to you its construction, you could easily understand that we use glass legs because these are capable of preventing the electricity from going away to the earth. If, therefore, I were to stand on this stool, and receive the electricity through this conductor, I could give it to anything that I touched. [The lecturer stood upon the insulating stool, and placed himself in connection with the conductor of the machine.] Now I am electrified; I can feel my hair rising up, as the paper tassel did just now. Let us see whether I can succeed in lighting gas by touching the jet with my finger. [The lecturer brought his finger near a jet from which gas was issuing, when, after one or two attempts, the spark which came from his finger to the jet set fire to the gas.] You now see how it is that this power of electricity can be transferred from the matter in which it is generated, and conducted along wires and other bodies, and thus be made to serve new purposes, utterly unattainable by the powers we have spoken of on previous days; and you will not now be at a loss to bring this power of electricity into comparison with those which we have previously examined, and to-morrow we shall be able to go farther into the consideration of these transferable powers.

LECTURE VI

THE CORRELATION OF THE PHYSICAL FORCES

WE have frequently seen, during the course of
these lectures, that one of those powers or forces
of matter, of which I have written the names on
that board, has produced results which are due to the action
of some other force. Thus you have seen the force of
electricity acting in other ways than in attracting; you have
also seen it combine matters together or disunite them by
means of its action on the chemical force; and in this case,
therefore, you have an instance in which these two powers
are related. But we have other and deeper relations than
these; we have not merely to see how it is that one power
affects another—how the force of heat affects chemical affinity, and so forth, but we must try and comprehend what relation they bear to each other, and how these
powers may be changed one into the other; and it will
to-day require all my care, and your care too, to make this
clear to your minds. I shall be obliged to confine myself to
one or two instances, because to take in the whole extent
of this mutual relation and conversion of forces would surpass the human intellect.

In the first place, then, here is a piece of fine zinc foil,
and if I cut it into narrow strips and apply to it the power
of heat, admitting the contact of air at the same time, you
will find that it burns; and then, seeing that it burns, you
will be prepared to say that there is chemical action taking
place. You see all I have to do is to hold the piece of zinc
at the side of the flame, so as to let it get heated, and yet
to allow the air which is flowing into the flame from all
sides to have access to it; there is the piece of zinc burning
just like a piece of wood, only brighter. A part of the zinc
is going up into the air in the form of that white smoke, and

part is falling down on to the table. This, then, is the action of chemical affinity exerted between the zinc and the oxygen of the air. I will show you what a curious kind of affinity this is by an experiment which is rather striking when seen for the first time. I have here some iron filings and gunpowder, and will mix them carefully together, with as little rough handling as possible; now we will compare the combustibility, so to speak, of the two. I will pour some spirit of wine into a basin and set it on fire; and, having our flame, I will drop this mixture of iron filings and gunpowder through it, so that both sets of particles will have an equal chance of burning. And now tell me which of them it is that burns? You see a plentiful combustion of the iron filings; but I want you to observe that, though they have equal chances of burning, we shall find that by far the greater part of the gunpowder remains untouched; I have only to drain off this spirit of wine, and let the powder which has gone through the flame dry, which it will do in a few minutes, and I will then test it with a lighted match. So ready is the iron to burn, that it takes, under certain circumstances, even less time to catch fire than gunpowder. [As soon as the gunpowder was dry, Mr. Anderson handed it to the lecturer, who applied a lighted match to it, when a sudden flash showed how large a proportion of gunpowder had escaped combustion when falling through the flame of alcohol.]

These are all cases of chemical affinity, and I show them to make you understand that we are about to enter upon the consideration of a strange kind of chemical affinity, and then to see how far we are enabled to convert this force of affinity into electricity or magnetism, or any other of the forces which we have discussed. Here is some zinc (I keep to the metal zinc, as it is very useful for our purpose), and I can produce hydrogen gas by putting the zinc and sulphuric acid together, as they are in that retort; there you see the mixture which gives us hydrogen—the zinc is pulling the water to pieces and setting free hydrogen gas. Now we have learned by experience that if a little mercury is spread over that zinc, it does not *take away* its power of decomposing the water, but *modifies* it most curiously. See how that mixture is now boiling; but when I add a little mercury to it the gas ceases to come off. We have now scarcely a bubble of hydrogen set free, so that the action is suspended for the time. We have not *destroyed* the power of chemical affinity, but modified it in a wonderful and beautiful manner. Here are some pieces of zinc covered with mercury exactly in the same way as the zinc in that retort is covered; and if I put this plate into sulphuric acid I get no gas, but this most extraordinary thing occurs, that if I introduce along with the zinc another metal which is *not* so combustible, then I reproduce all the action. I am now going to put to the amalgamated zinc in this retort some portions of copper wire (copper not being so combustible a metal as the zinc), and observe how I get hydrogen again, as in the first instance; there, the bubbles are coming over through the pneumatic trough, and ascending faster and faster in the jar; the zinc *now* is acting by reason of its contact with the copper.

Every step we are now taking brings us to a knowledge of new phenomena. That hydrogen which you now see coming off so abundantly does not come from the zinc, as it did before, *but from the copper.* Here is a jar containing a solution of copper. If I put a piece of this amalgamated zinc into it, and leave it there, it has scarcely any action; and here is a plate of platinum which I will immerse in the same solution, and might leave it there for hours, days, months, or even years, and no action would take place; but, by putting them both together, and allowing them to touch (FIG. 44), you see what a coating of copper there is immediately thrown down on the platinum. Why is this? The platinum has no power of itself to reduce that metal from that fluid, but it has, in some mysterious way, received this power by its contact with the metal zinc. Here, then, you see a strange transfer of chemical force from one metal to another; the chemical force from the zinc is transferred and made over to the platinum by the mere association of the two metals. I might take, instead of the platinum, a piece of copper or of silver, and it would have no action of its own on this solution, but the moment the zinc was introduced and touched the other metal, then the action would take place, and it would become covered with copper. Now is

not this most wonderful and beautiful to see? We still have
the identical chemical force of the particles of zinc acting,
and yet, in some strange manner, we have power to make
that chemical force, or something it produces, travel from

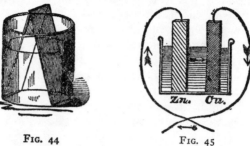

FIG. 44   FIG. 45

one place to another; for we do make the chemical force
travel from the zinc to the platinum by this very curious
experiment of using the two metals in the same fluid in
contact with each other.

Let us now examine these phenomena a little more closely.
Here is a drawing (FIG. 45) in which I have represented a
vessel containing the acid liquid and the slips of zinc and
platinum or copper, and I have shown them touching each
other *outside* by means of a wire coming from each of
them (for it matters not whether they touch in the fluid or
outside; by pieces of metal attached, they still, by that
communication between them, have this power transferred
from one to the other). Now if, instead of only using one
vessel, as I have shown there, I take another, and another,
and put in zinc and platinum, zinc and platinum, zinc and
platinum, and connect the platinum of one vessel with
the zinc of another, the platinum of this vessel with the zinc
of that, and so on, we should only be using a series of these
vessels instead of one. This we have done in that arrangement which you see behind me. I am using what we call a
Grove's voltaic battery, in which one metal is zinc and the
other platinum; and I have as many as forty pairs of these
plates all exercising their force at once in sending the whole
amount of chemical power there evolved through these wires
under the floor and up to these two rods coming through the

table. We need do no more than just bring these two ends in contact, when the spark shows us what power is present; and what a strange thing it is to see that this force is brought away from the battery behind me, and carried along through these wires! I have here an apparatus (FIG. 46) which Sir Humphrey Davy constructed many years ago, in order to see whether this power from the voltaic battery caused bodies to attract each other in the same manner as the ordinary electricity did. He made it in order to experiment with his large voltaic battery, which was the most powerful then in existence. You see there are in this glass jar two leaves of gold, which I can cause to move to and fro by this rack-work. I will connect each of these gold leaves with separate ends of this battery, and if I have a sufficient number of plates in the battery, I shall be able to show you that there

FIG. 46

will be some attraction between those leaves even before they come in contact; if I bring them sufficiently near when they are in communication with the ends of the battery, they will be drawn gently together; and you will know when this takes place, because the power will cause the gold leaves to burn away, which they could only do when they touched each other. Now I am going to cause these two leaves of gold to approach gradually, and I have no doubt that some of you will see that they approach before they burn, and those who are too far off to see them approach will see by their burning that they have come together. Now they are attracting each other, long before the connection is complete, and there they go! burnt up in that brilliant flash, so strong is the force. You thus see, from the attractive force at the two ends of this battery that these are really and truly electrical phenomena.

Now let us consider what is this spark. I take these two ends and bring them together, and there I get this glorious spark like the sunlight in the heavens above us.

What is this? It is the same thing which you saw when I discharged the large electrical machine, when you saw one single bright flash; it is the same thing, only *continued,* because here we have a more effective arrangement. Instead of having a machine which we are obliged to turn for a long time together, we have here a *chemical* power which sends forth the spark; and it is wonderful and beautiful to see how this spark is carried about through these wires. I want you to perceive, if possible, that this very spark and the heat it produces (for there is heat) is neither more nor less than the chemical force of the zinc—its *very* force carried along wires and conveyed to this place. I am about to take a portion of the zinc and burn it in oxygen gas for the sake of showing you the kind of light produced by the actual combustion in oxygen gas of some of this metal. [A tassel of zinc-foil was ignited at a spirit lamp and introduced into a jar of oxygen, when it burnt with a brilliant light.] That shows you what the affinity is when we come to consider it in its energy and power. And the zinc is being burned in the battery behind me at a much more rapid rate than you see in that jar, because the zinc is there dissolving and *burning,* and produces here this great electric light. That very same power which in that jar you saw evolved from the actual combustion of the zinc in oxygen, is carried along these wires and made evident here; and you may, if you please, consider that the zinc is burning in those cells, and that *this* is the light of that burning [bringing the two poles in contact and showing the electric light]; and we might so arrange our apparatus as to show that the amounts of power evolved in either case are identical. Having thus obtained power over the chemical force, how wonderfully we are able to convey it from place to place! When we use gunpowder for explosive purposes, we can send into the mine chemical affinity by means of this electricity; not having provided fire beforehand, we can send it in at the moment we require it. Now here (FIG. 47) is a vessel containing two charcoal points, and I bring it forward as an illustration of the wonderful power of conveying this force from place to place. I have merely to connect these by means of wires to the opposite ends of the

battery, and bring the points in contact.   See what an
exhibition of force we have!   We have exhausted the air
so that the charcoal can not burn, and therefore the light
you see is really the burning of the zinc in the cells behind
me; there is no disappearance of the
carbon, although we have that glorious electric light; and the moment
I cut off the connection it stops.
Here is a better instance to enable
some of you to see the certainty
with which we can convey this
force, where under ordinary circumstances, chemical affinity would not
act.   We may absolutely take these
two charcoal poles down under
water, and get our electric light
there.   There they are in the water, and you observe, when I bring
them into connection, we have the
same light as we had in that glass
vessel.

Fig. 47

Now besides this production of
light, we have all the other effects and powers of burning zinc.
I have a few wires here which are not combustible, and I am
going to take one of them, a small platinum wire, and suspend
it between these two rods which are connected with the

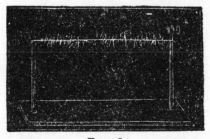

Fig. 48

battery, and when contact is made at the battery see what
heat we get (Fig. 48).   Is not that beautiful?   It is a
complete bridge of power.   There is metallic connection

all the way round in this arrangement, and where I have inserted the platinum, which offers some resistance to the passage of the force, you see what an amount of heat is evolved; this is the heat which the zinc would give if burnt in oxygen; but, as it is being burnt in the voltaic battery, it is giving it out at this spot. I will now shorten this wire for the sake of showing you that, the shorter the obstructing wire is, the more and more intense is the heat, until at last our platinum is fused and falls down, breaking off the circuit.

Here is another instance. I will take a piece of the metal silver, and place it on charcoal connected with one end of the battery, and lower the other charcoal pole on to it. See how brilliantly it burns! (FIG. 49.) Here is a

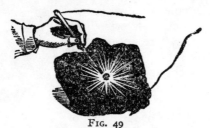

FIG. 49

piece of iron on the charcoal: see what a combustion is going on; and we might go on in this way, burning almost every thing we place between the poles. Now I want to show you that this power is still chemical affinity; that if we call the power which is evolved at this point *heat,* or *electricity,* or any other name referring to its source, or the way in which it travels, we still shall find it to be chemical action. Here is a colored liquid which can show by its change of color the effects of chemical action; I will pour part of it into this glass, and you will find that these wires have a very strong action. I am not going to show you any effects of combustion or heat, but I will take these two platinum plates, and fasten one to the one pole and the other to the other end, and place them in this solution, and in a very short time you will see the blue color will be entirely destroyed. See, it is colorless now! I have merely brought the end of the wires into the solution of indigo, and the power of electricity has come through these wires and made itself evident by its chemical action. There is also another curious thing to be noticed now we are dealing with the chemistry of electricity, which is, that the chemical power which destroys the color is only due to the action on one side. I will pour some more of this sulphindigotic acid([22]) into a flat dish, and will then make a porous dike of sand separating the two portions of fluid into two parts (FIG. 50), and now we shall be able to see whether there is any difference in the two ends of the battery, and which it is that possesses this peculiar action. You see it is the one on my right hand which

has the power of destroying the blue, for the portion on that side is thoroughly bleached, while nothing has apparently occurred on the other side. I say *apparently,* for you must not imagine that because you can not perceive any action none has taken place.

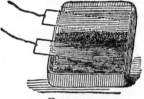

FIG. 50

Here we have another instance of chemical action. I take these platinum plates again and immerse them in this solution of copper, from which we formerly precipitated some of the metal, when the platinum and zinc were both put in it together. You see that these two platinum plates have no chemical action of any kind; they might remain in the solution as long as I liked, without having any power of themselves to reduce the copper; but the moment I bring the two poles of the battery in contact with them, the chemical action which is there transformed into electricity and carried along the wires again becomes chemical action at the two platinum poles, and now we shall have the power appearing on the left-hand side, and throwing down the copper in the metallic state on the platinum plate; and in this way I might give you many instances of the extraordinary way in which this chemical

[22] *Sulphindigotic acid.* A mixture of one part of indigo and fifteen parts of concentrated oil of vitriol. It is bleached on the side at which hydrogen gas is evolved in consequence of the liberated hydrogen withdrawing oxygen from the indigo, thereby forming a colorless deoxidized indigo. In making the experiment, only enough of the sulphindigotic acid must be added to give the water a decided blue color.

action or electricity may be carried about. That strange nugget of gold, of which there is a model in the other room, and which has an interest of its own in the natural history of gold, and which came from Ballarat, and was worth £8,000 or £9,000 when it was melted down last November, was brought together in the bowels of the earth, perhaps ages and ages ago, by some such power as this. And there is also another beautiful result dependent upon chemical affinity in that fine lead-tree([23]), the lead growing and growing by virtue of this power. The lead and the zinc are combined together in a little voltaic arrangement in a manner far more important than the powerful one you see here, because in nature these minute actions are going on forever, and are of great and wonderful importance in the precipitation of metals and formation of mineral veins, and so forth. These actions are not for a limited time, like my battery here, but they act forever in small degrees, accumulating more and more of the results.

I have here given you all the illustrations that time will permit me to show you of chemical affinity producing electricity, and electricity again becoming chemical affinity. Let that suffice for the present; and now let us go a little deeper into the subject of this chemical force, or this electricity—which shall I name first?—the one producing the other in a variety of ways. These forces are also wonderful in their power of producing another of the forces we have been considering, namely, that of magnetism; and you know that it is only of late years, and long since I was born, that the discovery of the relations of these two forces of electricity and chemical affinity to produce magnetism have become known. Philosophers had been suspecting this affinity for a long time, and had long had great hopes of success; for in the pursuit of science we first start with hopes and expectations; these we realize and establish, never again to be lost, and upon them we found new ex-

[23] *Lead tree.* To make a lead tree, pass a bundle of brass wires through the cork of a bottle, and fasten a plate of zinc round them just as they issue from the cork, so that the zinc may be in contact with every one of the wires. Make the wires to diverge so as to form a sort of cone, and, having filled the bottle quite full of a solution of sugar of lead, insert the wires and cork, and seal it down, so as to perfectly exclude the air. In a short time the metallic lead will begin to crystallize around the divergent wires, and form a beautiful object.

pectations of farther discoveries, and so go on pursuing, realizing, establishing, and founding new hopes again and again.

Now observe this: here is a piece of wire which I am about to make into a bridge of force, that is to say, a communicator between the two ends of the battery. It is copper wire only, and is therefore not magnetic of itself. We will examine this wire with our magnetic needle (FIG. 51), and, though connected with one extreme end of the battery, you see that before the circuit is completed it has

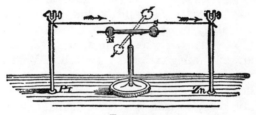

FIG. 51

no power over the magnet. But observe it when I make contact; watch the needle; see how it is swung round; and notice how indifferent it becomes if I break contact again; so, you see, we have this wire evidently affecting the magnetic needle under these circumstances. Let me show you that a little more strongly. I have here a quantity of wire which has been wound into a spiral, and this will affect the magnetic needle in a very curious manner, because, owing to its shape, it will act very like a real magnet. The copper spiral has no power over that magnetic needle at present; but if I cause the electric current to circulate through it, by bringing the two ends of the battery in contact with the ends of the wire which forms the spiral, what will happen? Why, one end of the needle is most powerfully drawn to it; and if I take the other end of the needle, it is repelled; so, you see, I have produced exactly the same phenomena as I had with the bar magnet, one end attracting and the other repelling. Is not this, then, curious to see, that we can construct a magnet of copper? Furthermore, if I take an iron bar, and put it inside the coil, so long as there is no electric current circulating round, it has no attraction, as you will observe if I bring a little iron filings or nails near the iron. But now, if I make contact with the battery, they are attracted at once. It becomes at once a powerful magnet, so much so that I should not wonder if these magnetic needles on different parts of the table pointed to it. And I will show you, by another experiment, what an attraction it has. This piece and that piece of iron, and many other pieces, are now strongly attracted (Fig. 52); but, as soon as I break contact, the power is all gone, and they fall. What, then, can be a better or a stronger proof than this of the relation of the powers of magnetism and electricity? Again: here is a little piece of iron which is not magnetized. It will not, at present, take up any one of these nails; but I will take a piece of wire and coil it round the iron (the wire being covered with cotton in every part it does not touch the iron), so that the current must go round in this spiral coil; I am, in fact, preparing an *electro-magnet* (we are obliged to use such terms to express our meaning, because it is a magnet made by electricity—because we produce by the force of electricity a magnet of far greater power than a permanent steel one). It is now completed, and I will repeat the experiment which you saw the other day, of building up a bridge of iron nails: the con-

Fig. 52

Fig. 53

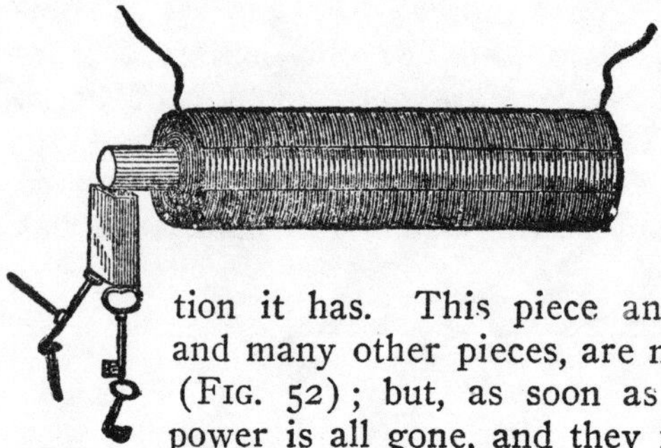

tact is now made and the current is going through; it is now a powerful magnet; here are the iron nails which we had the other day, and now I have brought this magnet near them they are clinging so hard that I can scarcely move them with my hand (Fig. 53). But when the contact is broken, see how they fall. What can show you better than such an experiment as this the magnetic attraction with which we have endowed these portions of iron? Here, again, is a fine illustration of this strong power of magnetism. It is a magnet of the same sort as the one you have just seen. I am about to make the current of electricity pass through the wires which are round this iron for the purpose of showing you what powerful effects we get. Here are the poles of the magnet; and let us place on one of them this long bar of iron. You see, as soon as contact is made, how it rises in position (FIG. 54); and if I take such a piece as this cylinder,

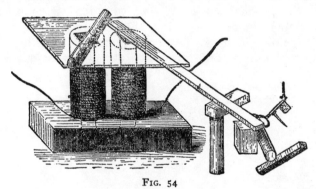

FIG. 54

and place it on, woe be to me if I get my finger between; I can roll it over, but if I try to pull it off, I might lift up the whole magnet, but I have no power to overcome the magnetic power which is here evident. I might give you an infinity of illustrations of this high magnetic power. There is that long bar of iron held out, and I have no doubt that if I were to examine the other end I should find that it was a magnet. See what power it must have to support not only these nails, but all those lumps of iron hanging on to the end. What, then, can surpass these evidences of the change of chemical force into electricity, and electricity into magnetism? I might show you many other experiments whereby I could obtain electricity and chemical action, heat and light from a magnet, but what more need I show you to prove the universal correlation of the physical forces of matter, and their mutual conversion one into another?

And now let us give place as juveniles to the respect we owe to our elders, and for a time let me address myself to those of our seniors who have honored me with their presence during these lectures. I wish to claim this moment for the purpose of tendering our thanks to them, and my thanks to you all for the way in which you have borne the inconvenience that I at first subjected you to. I hope that the insight which you have here gained into some of the laws by which the universe is governed, may be the occasion of some among you turning your attention to these subjects; for what study is there more fitted to the mind of man than that of the physical sciences? And what is there more capable of giving him an insight into the actions of those laws, a knowledge of which gives interest to the most trifling phenomenon of nature, and makes the observing student find

"Tongues in trees, books in the running brooks,
Sermons in stones, and good in every thing?"

ORDER FORM

GREAT BOOKS IN PHILOSOPHY PAPERBACK SERIES

ETHICS

Aristotle—*The Nicomachean Ethics*	$8.95
Marcus Aurelius—*Meditations*	5.95
Jeremy Bentham—*The Principles of Morals and Legislation*	8.95
Epictetus—*Enchiridion*	3.95
Immanuel Kant—*Fundamental Principles of the Metaphysic of Morals*	4.95
John Stuart Mill—*Utilitarianism*	4.95
George Edward Moore—*Principia Ethica*	8.95
Friedrich Nietzsche—*Beyond Good and Evil*	8.95
Bertrand Russell On Ethics, Sex, and Marriage (edited by Al Seckel)	17.95
Benedict de Spinoza—*Ethics* and *The Improvement of the Understanding*	9.95

SOCIAL AND POLITICAL PHILOSOPHY

Aristotle—*The Politics*	7.95
The Basic Bakunin: Writings, 1869–1871 (translated and edited by Robert M. Cutler)	10.95
Edmund Burke—*Reflections on the Revolution in France*	7.95
John Dewey—*Freedom and Culture*	10.95
G. W. F. Hegel—*The Philosophy of History*	9.95
Thomas Hobbes—*The Leviathan*	7.95
Sidney Hook—*Paradoxes of Freedom*	9.95
Sidney Hook—*Reason, Social Myths, and Democracy*	11.95
John Locke—*Second Treatise on Civil Government*	4.95
Niccolo Machiavelli—*The Prince*	4.95
Karl Marx/Frederick Engels—*The Economic and Philosophic Manuscripts of 1844* and *The Communist Manifesto*	6.95
John Stuart Mill—*Considerations on Representative Government*	6.95
John Stuart Mill—*On Liberty*	4.95
John Stuart Mill—*On Socialism*	7.95
John Stuart Mill—*The Subjection of Women*	4.95
Thomas Paine—*Rights of Man*	7.95
Plato—*The Republic*	9.95
Plato on Homosexuality: Lysis, Phaedrus, and *Symposium*	6.95
Jean-Jacques Rousseau—*The Social Contract*	5.95
Mary Wollstonecraft—*A Vindication of the Rights of Women*	6.95

METAPHYSICS/EPISTEMOLOGY

Aristotle—*De Anima*   6.95
Aristotle—*The Metaphysics*   9.95
George Berkeley—*Three Dialogues Between Hylas and Philonous*   4.95
René Descartes—*Discourse on Method* and *The Meditations*   6.95
John Dewey—*How We Think*   10.95
*The Essential Epicurus: Letters, Principal Doctrines, Vatican Sayings,
    and Fragments* (translated, and with an introduction,
    by Eugene O'Connor)   5.95
Sidney Hook—*The Quest for Being*   11.95
David Hume—*An Enquiry Concerning Human Understanding*   4.95
David Hume—*Treatise of Human Nature*   9.95
William James—*Pragmatism*   7.95
Immanuel Kant—*Critique of Pure Reason*   9.95
Gottfried Wilhelm Leibniz—*Discourse on Method* and the *Monadology*   6.95
Plato—*The Euthyphro, Apology, Crito,* and *Phaedo*   5.95
Bertrand Russell—*The Problems of Philosophy*   8.95
Sextus Empiricus—*Outlines of Pyrrhonism*   8.95

PHILOSOPHY OF RELIGION

Ludwig Feuerbach—*The Essence of Christianity*   8.95
David Hume—*Dialogues Concerning Natural Religion*   5.95
John Locke—*A Letter Concerning Toleration*   4.95
Thomas Paine—*The Age of Reason*   13.95
Bertrand Russell On God and Religion (edited by Al Seckel)   17.95

ESTHETICS

Aristotle—*The Poetics*   5.95

GREAT MINDS PAPERBACK SERIES

ECONOMICS

Adam Smith—*Wealth of Nations*   9.95

RELIGION

Thomas Henry Huxley—*Agnosticism and Christianity*
    and Other Essays   10.95
Ernest Renan—*The Life of Jesus*   11.95
Andrew D. White—*A History of the Warfare of Science*
    with Theology in Christendom   19.95

SCIENCE

Charles Darwin—*The Origin of Species*   10.95
Michael Faraday—*The Forces of Matter*   8.95
Galileo Galilei—*Dialogues Concerning Two New Sciences*   9.95
Ernst Haeckel—*The Riddle of the Universe*   10.95
Julian Huxley—*Evolutionary Humanism*   10.95

HISTORY

Edward Gibbon—*On Christianity*   9.95
Herodotus—*The History*   13.95

SPECIAL—For your library . . . the entire collection of 51 "Great Books in Philosophy" and 11 "Great Minds" available at a savings of more than 15%. Only $356.00 for the "Great Books" and $108.00 for the "Great Minds" (plus $12.00 postage and handling). Please indicate "Great Books/Great Minds—Complete Set" on your order form.

The books listed can be obtained from your book dealer or directly from Prometheus Books. Please indicate the appropriate titles. Remittance must accompany all orders from individuals. Please include $3.50 postage and handling for the first book and $1.75 for each additional title (maximum $12.00, NYS residents please add applicable sales tax). Books will be shipped fourth-class book post. **Prices subject to change without notice.**

Send to _____
(Please type or print clearly)

Address _____

City _____ State _____ Zip _____

Amount enclosed _____

Charge my ☐ VISA ☐ **MasterCard**

Account # ☐☐☐☐☐☐☐☐☐☐☐☐☐☐☐☐

Exp. Date _____/_____ Tel.# _____

Signature _____

Prometheus Books Editorial Offices
700 E. Amherst St., Buffalo, New York 14215

Distribution Facilities
59 John Glenn Drive, Buffalo, New York 14228

Phone Orders call toll free: (800) 421-0351
FAX: (716) 691-0137
Please allow 3-6 weeks for delivery